Gustave Mosselman

Manual of veterinary microbiology

Gustave Mosselman

Manual of veterinary microbiology

ISBN/EAN: 9783337215088

Printed in Europe, USA, Canada, Australia, Japan

Cover: Foto ©berggeist007 / pixelio.de

More available books at **www.hansebooks.com**

MANUAL

OF

VETERINARY MICROBIOLOGY.

BY

PROFESSORS MOSSELMAN AND LIENAUX,

National Veterinary College, Cureghem, Belgium.

TRANSLATED AND EDITED BY

R. R. DINWIDDIE,

Professor of Veterinary Science, College of Agriculture, Arkansas State University; Animal Pathologist, Arkansas Agricultural Experiment Station.

CINCINNATI:
THE ROBERT CLARKE COMPANY.
1894.

PREFACE.

In undertaking the translation of MM. Mosselman and Liénaux's "*Manuel de Microbiologie Vétérinaire*," my object has been to supply English speaking veterinary students and practitioners with a work on Bacteriology which seemed specially adapted to their needs. The book is small, but it conveys more information on the etiology of the infectious diseases of animals and the biology of the germs associated with them than any other single work in our language. Unlike other works on Bacteriology accessible to readers of English only, the Microbiology of animal diseases is treated of as the essential part of the work, that of diseases of mankind only incidentally referred to. The completeness and accuracy as to details with which it discusses the modes of propagation of some of our most important diseases and the general conditions under which these diseases occur, ought to recommend the book to practical veterinarians, who are presumably more interested in the ascertained facts in regard to any disease than in the individuality of the germ which occasions it.

The book is not intended for a laboratory manual, consequently, the technique of staining and cultivation of germs is not exhaustively discussed, and the usual illustrations of bacteriological apparatus have been omitted.

As to the translation itself a few words are necessary: Weights and measures, given in the metrical system in the original, have not been changed. Dimensions which occur in this work are chiefly those of microscopic objects which are now rarely expressed by American microscopists in fractions of an inch. The thermometric readings are in all cases given in the centigrade scale. For those who are unfamiliar with the decimal system the Appendix will supply the requisite information.

The few foot-notes which I have introduced are in some cases intended to be supplementary to the text, referring to discoveries which have been made since the publication of the original in 1891. In other cases they are explanatory of words or statements which might otherwise be misunderstood.

To the illustrations which occur in the French text I have added a few others borrowed from different sources, which are acknowledged in the descriptions accompanying the figures; four are from drawings of preparations in my own possession.

R. R. Dinwiddie.

Fayetteville, Ark., *July* 19, 1894.

TABLE OF CONTENTS.

PART FIRST.

Generalities upon Microbes.

PART SECOND.

Generalities upon Pathogenic Microbes.

PART THIRD.

Microbic Diseases Individually Considered.

INTRODUCTION.

Works which treat of Microbiology are quite numerous, but none offer a concise and complete exposition of the accepted facts on the subject, the application of which is within the reach of students and practitioners. Such outline we would give here in the hope that both may be benefited thereby.

It would seem that this publication has some prospect of being well received. Besides the fact that the veterinarian in daily practice is under the necessity of having recourse to the teachings of microbiology, the inspection of meat and the supervision of the sanitary police—duties which have devolved upon him—make it his imperative duty to neglect no means of diagnosis which science places at his disposal.

We do not mean to assert that the diagnosis of infectious diseases necessitates in all cases a search for the pathogenic microbes, but, recognizing the importance of the pathological anatomy and clinical symptoms, we believe that the demonstration of these germs is of much higher value. We will even say that, in unfortunately too many cases, the recognition of the germs is the only mode of definitely establishing the nature of a lesion.

Every one can understand how desirable it is that the practitioner, meat inspector or sanitary veterinarian, whose decisions very frequently run counter to some particular interest, should pronounce himself only after having made use of this last resource which will protect him from scientific mistakes and contradictions, always much to be regretted.

A like exactitude is to be desired in private practice, where it will form a basis for sound therapeutics.

We will briefly trace the history of those microbes which may be of interest to the practitioner, and describe the technique of those investigations which he may daily be called upon to make. Our study will embrace three divisions. In the first we will briefly consider the subject of microbes in general; in the second we will study pathogenic germs collectively; and in the third notice those particular microbes which occasion disease in animals and even in mankind.

MANUAL
OF VETERINARY MICROBIOLOGY.

PART FIRST.

MICROBES CONSIDERED IN GENERAL.

Microbes may be considered successively in the static and in the functional or physiological condition.

CHAPTER I.

MICROBES IN THE STATIC CONDITION.

Definition—1. Forms of microbes; 2. Organization and chemical composition; 3. Situation and distribution.

The names *Microbes*, *Bacteria*, *Vibrios*, *Schizomycetes*, *Schizophytes*, have been given to unicellular microscopic beings placed at the bottom of the scale of the vegetable kingdom. These beings, destitute of chlorophyll, live at the expense of complex organic substances, which they reduce to the condition of simple mineral compounds.

I. *Forms of microbes.*

1. *Typical forms.*—The form of microbes is that of a rounded corpuscle or of a rod. The latter may be straight, undulated, or spiral.

Rounded microbes or *cocci* have received the following names, according to the manner in which they are grouped:

Fig. I.

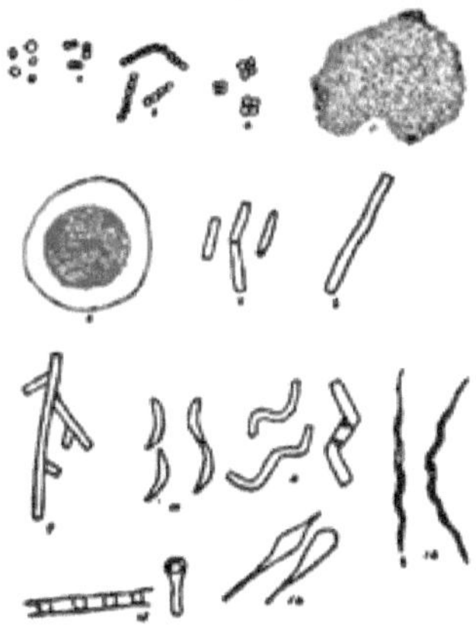

Micrococcus: cocci isolated, (Fig. I, 1);

Diplococcus: cocci arranged in pairs, (2);

Streptococcus: cocci arranged in linear series, in chains, (3);

Micrococcus tetragenus: cocci arranged in groups of four, (4);

Sarcina: cocci arranged in tetrads so as collectively to form a cube.

Staphylococcus: cocci associated in clusters;

Zoöglœa: cocci associated in large numbers in an amorphous matrix, (5);

Ascococcus: cocci associated in large numbers in an amorphous matrix and inclosed in an enveloping membrane, (6).

The elongated microbes are cylindrical or fusiform rods, or have the shape of a bell clapper. They have received the following names:

Bacillus: rods short and straight, (7);

Leptothrix: rods long and undulating, (8);

Cladothrix: rods long, straight, and branching, (9).

Spiral microbes are in the form of an arc of a circle, or are spiral. They appear, however, with rectilineal forms when their curvature is directed toward the objective. The following terms are used to designate them:

Vibrio: microbes spiral, short, (10);
Spirillum: microbes spiral, long and rigid, (11);
Spirochæte: microbes spiral, long and flexible, (12).

2. *Forms of involution, of degeneration.*—Under special conditions unfavorable to their nutrition bacteria may assume abnormal aspects, such as swelling in the form of a club, at which place the protoplasm becomes clearer: these special forms are called forms of involution, (14). They have been established in the actinomyces, the bacillus of Koch, etc.

3. *Polymorphism.*—Microbes are essentially polymorphic. Recent researches have shown that the same microbe may assume very different aspects, according to the medium in which it lives. Thus, the germ of the pyocyanic disease presents itself successively as a bacillus, a spirillum, and a micrococcus. The bacillus of Pasteur's septicemia grows in long filaments in the blood, in short bacilli in the subcutaneous cellular tissue. The bacillus of symptomatic charbon, cultivated in bouillon containing glycerin and sulfate of iron, takes the form of a clove. With the same germ, therefore, we can obtain several morphologically distinct individualities.

The dimensions of microbes are as variable as their form; in all cases these are expressed by a few thousandths or even fractions of the thousandth of a millimeter.*

* [The dimensions of microscopic objects are usually expressed in *Microns.* A micron is the one thousandth part of a millimeter, and is designated by the Greek letter μ. The dimensions of microbes, expressed in the original of this work in decimal fractions of a millimeter, have been rendered in the translation as microns. Thus Omm., 005=5μ.—D.]

II. *Organization of microbes.*

Structure and chemical composition.—The structure of microbes may be compared to that of a cell without nucleus. The existence of the latter is not generally admitted, although some authorities believe they have observed such a structure.

The content is a kind of protoplasm which Nenki has designated by the name of mycoprotein, a homogeneous, or sometimes granular, substance generally devoid of color. Under certain circumstances it may contain brilliant corpuscles (spores), starch grains (in the Sarcina), and granules of sulphur (Beggiotoa).

The periphery is formed by a thin and flexible, or thick membrane, the nature of which is not well known. Most authors look upon it as a carbo-hydrate allied to cellulose; its resistance to acids and alkalies seems to support this view. Others regard it as a layer of mycoprotein differentiated from the protoplasm.

Sometimes this membrane emits vibratile cilia; at other times it is surrounded by a zone of a mucilaginous aspect, capable of swelling up in water and forming a transparent capsule to the germ (pneumococcus).

III. *Situation and distribution of germs in nature.*

The part which these beings play in nature being known, it is easy to understand that they should be found wherever there is organic matter to be reduced. We find them, indeed, pullulating in all places where there are no special conditions prejudicial to their life. We will notice their distribution in the most important media: atmospheric air, waters, soil, foods, the living organism, dwellings, vehicles, clothing, etc.

Air.—The germs which are found suspended in the atmosphere can not multiply there, and hence come from other media from which they are carried off by atmospheric currents along with pulverulent matters. They are deposited in the calm (tranquil air, hollow places). Their number and nature vary with climatic and other conditions; their number increases during desiccation of the soil (summer), and diminishes after rains. They are most numerous in inhabited places and in the vicinity of marshes, whilst the air of mountains and of the surface of seas is almost completely free from them.

The germs of the air do not long resist the combined action of oxygen and light; nevertheless, through the agency of the winds their effects may be manifested at great distances.

The study of atmospheric germs is made by simple enumeration with the microscope or by various culture processes. The latter method is much to be preferred since it allows of the separation of dead germs, the number of which is very large.

Waters.— Subterranean waters, having filtered through thick layers of earth, are free from all germs, but quickly become infected on contact with the surface soil and the air. The waters of wells are always infected; their pollution is, moreover, easy to understand: the masonry having no support at the bottom eventually sinks down and fissures are produced through which infiltrate the waters of the neighboring surface soils often impregnated with germs. For this reason wells should not be built in proximity to cisterns, cess-pools or dung-hills.

The deep waters best protected against infection are those of artesian wells.

Surface waters are always very rich in germs, the nature of which is extremely variable; the majority are ubiquitous germs, pathogenic being much less frequent. Stagnant waters especially favor the multiplication of germs.

Soil.—Whilst the rocks and the virgin soil from the depths are free from all germs, these occur in large numbers in the superficial layers. Their number and nature vary infinitely according to location, season, winds, the physical constitution and chemical composition of the soils, etc. Their multiplication gradually decreases as the depth increases. Water, in filtering through the ground, yields up the germs which it contains as well as the soluble matters which serve for their nutrition.

It may be said, in truth, that germs in way of pullulation, through their power of penetrating the capillary spaces, should of themselves sink below the layer of soil in which the waters deposit them. This vegetation, however, is itself impeded by unfavorable conditions of temperature and nutrition, the absence of oxygen, etc. Hence in good filtering soil germs are no longer found at a depth of three meters.

Foods.—Vegetable foods (fodders, oats, etc.) are always contaminated with germs derived from the air, the soil, or waters. Animal foods are generally contaminated by contact with the air. Foods, whether of vegetable or animal origin, are especially favorable to the multiplication of microbes. The various means employed for the preservation of foods have no other aim than to protect them against the invasion or the destructive action of these organisms. When these means are defective or powerless to arrest the evolution of the germs which have been deposited there, various changes supervene which diminish the intrinsic nutritive value of the foods and may even render them detrimental to the health of man or of animals (damaged hay, putrid meat). The contamination of foods by pathogenic germs, properly so called, will be studied later.

Houses and vehicles.—The walls, floor, and ceiling, as well as the mangers and racks, of houses occupied by animals are constantly liable to receive the germs which are borne in the atmospheric dust, cleansing waters, solid dejections, litter, foods, etc.

Vehicles (wagons, etc.) serving for the transport of animals may be contaminated by microbes in the same way as houses.

Harness, blankets, tools, and other objects.—It is easy to understand that these objects will most frequently be contaminated either by the various methods mentioned above or by their contact with the animal for the use of which they are destined.

Organism.—After what we have said of the nutrition of germs, we may expect to encounter them in all parts of the economy which are in direct relation with the air, or with solid and liquid media.

The digestive canal throughout all its course contains in large numbers various microbic species which have been carried there by food and drink. In the mouth we find especially a leptothrix, spirochæte, and vibrios; from it have also been isolated pathogenic germs—staphylococcus pyogenes, etc.

The stomach contains especially sarcinae, yeasts, and elongated bacteria, whilst the intestines contain large numbers of bacilli. Micrococci and elongated non-sporulated bacteria are killed by contact with the gastric juice, and, consequently, do not multiply in the intestine.

The mucus of the anterior respiratory passage is likewise always contaminated with germs which have been deposited by the inspired air, the latter itself being thus purified so as to emerge free from all germs. Hence, it apppears that disease infection can not take place through the expired air. The ocular mucosa and genito-urinary mucosa near the external openings contain also a certain number of microbes.

Finally, these are found lodged upon the skin, the perspiration and sebaceous secretion along with the epidermic debris normally cast off constituting a good medium for their preservation.

The blood of healthy animals is free from germs. In the pathological condition, on the other hand, most of the tissues and fluids of the organism may become the seat of the evolution of bacteria.

CHAPTER II.

PHYSIOLOGY OF MICROBES.

1. Digestion.—2. Respiration.—3. Nutrition.—4. Movements.—5. Generation and multiplication.—6. Action of the media upon microbes.—7. Action of microbes upon the media.—8. Classification.

I. *Digestion.*

1. *Foods.*—Microbes being destitute of chlorophyll require for their nutrition organic products already formed; consequently they must nourish themselves at the expense of vegetable or animal substances. They borrow nitrogen from albuminoid substances or their derivatives as well as from ammoniacal salts, and occasionally, in part, from nitrates; carbon and hydrogen from hydrated carbonaceous substances—sugar, glycerin, and salts of malic, tartaric and acetic acids.

They require also mineral substances—sulfates and phosphates of sodium, potassium, and magnesium.

They are very sensitive to the chemical composition of the nutrient medium in which they live; traces of certain substances, as well as the absence of others, can bring about profound alterations in the manifestations of their vitality.

Their medium should be slightly alkaline and very aqueous. Excess of acidity or of alkalinity, acidity especially, is prejudicial to their growth.

2. *Digestion.*—The foods of microbes, like those of

animals and vegetables, independent of the chlorophyllic function of the latter, require to undergo certain modifications preparatory to assimilation. These modifications, representing the digestion of microbes, consist of a hydration accompanied or not by a splitting up of molecules.

This phenomenon is accomplished by means of soluble ferments secreted by the germs, and it is remarkable that these digestive ferments are the same as those which are found in higher beings; thus, in the case of microbes as for these last, starch is transformed into dextrin by a diastase[1] called *amylase* corresponding to vegetable diastase, ptyalin, and to the amylaceous ferment of the pancreatic juice. Cane sugar is split up into glucose and levulose by a diastase or *sucrase* identical with the invertin of the beet-root and with the inverting ferment of the intestinal juice; albuminoid substances are peptonized by microbes through the secretion by the latter of a special pepsin or casease; in the case of the casein of milk the action of this last substance is preceded by that of a diastase analogous to *rennet* which, like the latter, determines the coagulation of the milk.

When we consider the great number of germs contained within the digestive canal, the comparison which we have just made between digestion in microbes and in the higher beings, indicates the possibility of an adjuvant action of the former in the digestion of animals.

[1] The name *diastase*, formerly limited to the amylaceous ferment of vegetables, is now synonymous with soluble ferment, thus amylase, sucrase, pepsin, rennet (présure), are *diastases* or *zymases*.

These are not the only diastases secreted by microbes. We are far from knowing all of them; they vary, naturally, according to the special nature of their food. We have noted the principal of them in order to bring out the general mode of the nutritive process in microbes.

II. *Respiration.*

The study of the respiration of microbes is of great interest. Obviously, all require oxygen which, in oxidizing alimentary substances, supply the heat necessary for the maintenance of life, for multiplication and motion, etc. Many of them borrow it in the free state from the atmosphere or from water (*aërobes*), but there are others which appear incapable of enduring free oxygen, hence require to live protected from the air (*anaërobes*). These last act upon certain organic substances by a sort of internal combustion; they reduce these substances into carbonic acid and into other molecules generally less complex, but still susceptible of oxidation, setting free a certain amount of energy, as the aërobic germs on their part do by a true combustion.

One of the prominent characters of anaërobic germs when they are nourished at the expense of quaternary substances consists in the disengagement of abundant gaseous products, among which we find, besides carbonic acid, nitrogen, ammonia, and ammoniacal compounds (trimethylamin, etc.), sulfuretted and phosphoretted hydrogen, etc.; these mixtures emit a peculiarly fetid odor (putrid gases). When they especially reduce ternary products these anaërobes give as gases carbonic acid, hydrogen, and hydro-carbons.

A certain number of germs accommodate themselves equally well to both these ways of life; in the air they are aërobic, in its absence they become anaërobic. This double faculty has been expressed by the term *aëro-anaërobic.*

III. *Nutrition.*

1. *Absorption and assimilation.*—Absorption of food takes place by osmosis. A part of the principles absorbed is utilized for the elaboration of plastic material; another part behaves as a respiratory food. The formation of plastic material must be considerable when we take into account the excessively rapid multiplication of micro-germs.

The respiratory foods serve especially for the production of the different forms of work performed by the elements, and which are represented by the phenomena of assimilation, growth, locomotion, heat, sometimes light (phosphorescence of meat and fish).

Their intimate nutrition is little known. Chemically, the point of departure of the nutritive action is quite different according to the case. Some require albuminoids, whilst others draw their nitrogen from azotized products with molecules of much less complexity: leucin, tyrosin, xanthin, etc.; others again borrow it from trimethylamin and from ammoniacal salts. The same variety is observed in the case of non-nitrogenous foods.

This peculiarity accounts for the successive appearance of different bacteria in an organic medium abandoned to the external air. As this medium becomes more and more exhausted those species of germs successively appear whose lesser requirements permit of their living at the expense of the nutritive residue of those which have preceded them.

2. *Disassimilation, excretions, secretions.*—From the preceding considerations it results that disassimilation in microbes ought to give very varied residues. These residues naturally depend upon the food, upon the species of germ, and upon the special conditions in which their evolution is accomplished (temperature, aërobic or anaërobic nature, etc.).

Of these residual or excrementitial products of microbes some are *gaseous* (carbonic acid, hydrogen, carburetted hydrogen, sulfuretted hydrogen, ammonia), some *volatile* (trimethylamin, alcohol, formic, acetic, butyric acids, etc.), some *fixed* (lactic and malic acids, leucin, taurin, tyrosin, etc., etc.).

The nutrition of germs may give rise to coloring matters, such germs being called *chromogenic.* The coloring matter thus produced may be soluble or insoluble; in the former case it diffuses in the fluid media in which the germs occur, an instance of which may be seen in the germ of blue milk.

Among the number of the substances resulting from the nutrition of microbes we have to mention the *ptomaines.*

Ptomaines are ammoniacal compounds acting the part of bases, and which, upon the higher beings, have often effects analogous to those of the vegetable alkaloids, which they resemble in every respect.

In a general way the residual products are noxious to the germs from which they spring: 0.8 per cent of free butyric acid arrests the butyric fermentation of lactate of lime.

The *diastases* must also be cited among the products

of the nutrition of microbes; they are secreted for the requirements of digestion, as we have already seen.

Later, we shall have occasion to see, when considering the rôle of microbes, that their nutrition is the determining cause of the chemical reactions which characterize fermentations and putrefaction.

IV. *Movements of microbes.*

Some bacteria are immobile (the majority of round and some elongated bacteria), others are gifted with the faculty of moving themselves in the fluids in which they live.

The kind of movement varies with the species concerned; sometimes the element, maintaining its rectilineal direction, performs a simple, more or less regular oscillation around an imaginary longitudinal axis; at other times it undergoes a slight inflection in the direction of its length and straightens itself again alternately; at other times, again, it assumes a flexuous appearance simulating the movements of a snake; some even wind themselves around in corkscrew fashion.

The motion of a certain number of bacteria is dependent upon the presence of vibratile prolongations, in others these movements seem to depend upon contractions taking place within the body of the element.

In all they are directly dependent upon nutrition the integrity of which is necessary to their production.

Light and the fluidity of the media are conditions which favor them.

V. *Generation, multiplication.*

1. *Spontaneity.*—Formerly it was supposed that microbes originated by spontaneous generation in putrefactive media; this origin, indeed, was accepted for all beings the mode of reproduction of which was unknown. The progress of the natural sciences first considerably restricted the scope of this theory, which the experiments of Pasteur triumphantly and absolutely combatted. Although it can not be denied that at a period in the remote past organized matter must have been formed spontaneously at the expense of mineral matters, it seems well established now that the molecular association which tends to the constitution of protoplasm is no more produced, at least within the conditions of observation accessible to man, except at the expense of a pre-existent being. We therefore have to consider here, from a practical point of view, only the reproduction of germs by multiplication.

2. *Fission.*—Microbes multiply principally by fission. The cells of which they are composed become elongated, then divided into two by a transverse groove; the two segments which result from this division may separate and live independently or may remain united so as to form agglomerations of various kinds; for example: the chains or chaplets of micrococci which adhere end to end; the zoöglœa to which the same micrococci give rise when united in mass by a gelatinous substance, the jointed filaments of the anthrax bacillus, etc.

Fission usually takes place in one direction only,

but there are bacteria in which the division takes place in two crossed directions (micrococcus tetragenus), or even in three directions (sarcina); in this last case the bacteria, the secondary elements of which remain united, take the form of a cube.

3. *Sporulation.*—Multiplication by fission appears to be the only mode possessed by microbes of spherical form; in the majority of others we recognize a second mode,—sporulation (13). This consists in the formation within the bacteria of brilliant points which are apparently the result of a condensation of the original protoplasm, whilst the latter at the same time becomes very clear. These brilliant points are the spores; they are set at liberty by the destruction of the cell which has produced them and when they find themselves in good conditions of temperature and humidity, and in a suitable medium, they reproduce the bacteria as, in the higher forms of vegetation, the seed gives origin to the entire plant.

Spores show a remarkable resistance to the action of the common causes of destruction of microbes. They almost never develop in the media in which they have taken birth.

Botanists recognize, besides fission, which for them is only a form of growth, or vegetation, two methods of sporulation or fructification. The first and best known is *endosporulation* which we have just described; the second which it is not always easy to distinguish from fission has received the name *arthrosporulation*. It is characterized by the production by fission, at the expense of cells performing the function of reproductive elements, of new cells which differ from those obtained by ordinary fission by their state of latent life and by the thickness and resistance of their enveloping membrane. They are generally larger than the spores begotten by endosporulation and resemble cysts. It is this distinction in the mode of formation of spores which serves as the fundamental basis of the classification of Guignard which we will reproduce later.

It is not without interest to have an idea of the power of multiplication possessed by these micro-organisms. If we regard a bacterium as dividing itself into two after one hour, we will have four of them in two hours, and in twenty-four hours, 16,000,000. After forty-eight hours we will have the fabulous number of 280 trillions. We can estimate from this the ravages that germs introduced into the blood must produce when they find there conditions favorable to their pullulation.

VI. *Action of the media upon microbes.*

The medium has necessarily a great influence upon microbes. Besides supplying them with food it is capable of modifying their vitality in different degrees.

Certain agents or conditions have the power of bringing bacteria to the condition of latent life, of changing their usual mode of activity, or even of destroying them. These agents or conditions are of a nature either mechanical, physical, chemical, or physiological.

1. *Mechanical influences.*—These are badly determined; some authors claim to have observed that oscillations impressed upon cultures of microbes are

adverse to their multiplication; other experimenters have arrived at directly opposite results.

2. *Physical influences.—a) Humidity.*—Water is indispensable to microbes; a medium containing less than sixty per cent of this liquid arrests their multiplication. Desiccation causes all active life to cease (latent life) and may in this way finally lead to their death.

Similarly, it is to their deficiency in water content that we must attribute the preservation of vegetable juices by sugar, of meat pastry by fat, etc.

b) Temperature.—A temperature too high or too low is detrimental to microbes. They are generally more sensitive to an elevation of temperature than to a depression. The vegetative forms are killed by two hours' exposure to a temperature of 48° to 60° C. The spores, however, are much more resistant and require for their destruction a temperature of 140° when they are in a dry medium, 100° when they are moist.

The temperature of predilection of microbes is from 20° to 39°; above the latter temperature disturbances of microbic activity, either temporary or permanent, are liable to ensue (principle of attenuation by heat).

Cold arrests the multiplication of microbes (latent life), but kills them only with difficulty. Some have been exposed to a temperature of —105° without their vitality being completely destroyed.

c) Light.—Light is a puissant cause of destruction to microbes; it excites oxidation of their constituent principles and especially of the hydro-carbonaceous substances; its action is quite rapid and continues even after the germs are again removed from the light.

d) Electricity.—The action of continuous and of

induction currents have been studied. This action, however, is but little known; in the case of continuous currents it is necessary to take into account the electrolysis which separates the acids from the bases and transfers these new molecules to the two poles. The acid reaction of the positive pole is opposed to microbic pullulation at this point; the alkalinity of the negative pole is less energetic in its action. The current may act upon the germ itself and interfere with its multiplication, especially when strong currents are employed. It is not inadmissible to hope that, the influence of electricity becoming better known, its effects may some day be utilized in the treatment of infectious diseases.

3. *Chemical influences.*—The exhaustion of the nutritive medium, accumulation of residual products, excess of alkalinity and more especially of acidity, oppose and may even arrest the multiplication of germs.

Certain chemical substances exert toxic effects upon microbes, this toxicity varying with the microbic species concerned. The rational application of this action constitutes the basis of antisepsis from which hygiene and medicine have already derived great profit.

4. *Physiological influences.*—When several microbic species occur in the same place they may oppose each other and then a veritable struggle for existence ensues. From this concurrence may result the annihilation of the least favored species; the medium, the number of germs, and various other circumstances may intervene here. In putrefactive media in which

the bacterium termo and bacillus subtilis live we do not find the bacillus septicus.

Under other circumstances it may happen that the two associated germs assist each other, the one preparing alimentary materials for the other, or, it may be, protecting the other from the noxious action of certain agents (association of aërobic and anaërobic germs).

We ought to mention in this connection the influence exerted upon microbes by the tissue elements of animals in which they sometimes live (*phagocytosis*); we will enter more into the details of this subject in the second part.

VII. *Action of microbes upon the media.*

1. RÔLE OF THE BACTERIA IN NATURE.

From a general biological point of view the rôle of microbes consists in reducing to the condition of simple inorganic compounds the organic matter built up by vegetables and incorporated by animals. The disassimilation which occurs in these beings correlative with nutrition destroys a part of this organic matter and reduces it to the condition of carbonic acid, water, and salts. This constantly occurring decomposition, however, not only fails to completely mineralize the substance upon which it acts, but, in addition, at the death of animals and plants an amount of elaborated substance remains which, in future, is exempt from this cause of destruction. It is here that microbes or, more accurately, ferments enter on the scene.

"The life of the larger forms of vegetation builds

up in nature at the expense of the solar heat, substances whose production requires a certain expenditure of force. It is in these endothermic subtances that the lower organisms implant themselves. From the energy which they find there stored up they borrow a portion for the construction of their own tissues, which renders them up to a certain point independent of external conditions. Another portion is used to convert into the gaseous condition substances originally fluid or solid. Another, finally, is transformed into sensible heat and serves to elevate the temperature of the fluid in which all these phenomena occur, and, as a consequence, to accelerate their production." (Duclaux.)

To accomplish this immense work the ferments are endowed with an intense destructive power, and operate, thanks to the rapidity of their multiplication, in innumerable legions.

We must here briefly refer to fermentations and putrefaction.

Fermentations.—Fermentations are always the result of the intervention of micro-organisms. They consist in modifications of special organic substances tending to the formation of simpler products in which the heat of total combustion is less than that of the fermentable substances from which they are derived. The difference between these quantities of heat represents the amount of energy appropriated by the germ for its nutritive requirements, and the reaction by which the fermentation is characterized has no other object than the liberation of this energy.

Fermentable substances are comparatively few in number; they are usually bodies rich in oxygen—

carbo-hydrates, polyatomic alcohols, the lower fatty acids, and albuminoids. According to the substance predominating in their products, fermentations are described as alcoholic, acetic, lactic, butyric, viscous, ammoniacal, etc.

Microbes capable of developing fermentations are called *zymogenic.*

Putrefaction.—By putrefaction is meant the decomposition of the substance of organized beings through the agency of microbes. This decomposition supervenes shortly after death; exceptionally it is observed during life as we will see in the special part of this work. The microbes which determine it are called *septogenic.* It consists in a series of fermentations so associated that the products of the one form the point of departure for the following. These fermentations occur simultaneously and take effect upon the various immediate principles of the organism: albuminoid substances, hydrocarbonaceous bodies, etc.

But the decomposition of these last named principles giving rise to fermentations simpler and better known, and specially denominated as such, the term putrefaction refers more particularly to the microbic degradation of albuminoid molecules.

Many microbic species are concerned in putrefaction, these species varying according to the case; there are some which are quite frequently met with, such as *bacterium termo*, *bacillus subtilis* (hay bacillus), and *bacillus septicus*, micro-germs whose characteristics will be indicated later, we always meet with an association of aërobic and anaërobic organisms.

The bodies which arise during putrefaction are numerous: hydrogen, sulfuretted, carburetted and phosphoretted hydrogen; ammonia, carbonic, formic, acetic, butyric, and lactic acids, etc.; amines, trimethylamin, propylamin, etc.; indol, scatol, tyrosin, ptomaines, etc.

The term *saprogenic* is applied more particularly to those microbes which excite putrefaction with disengagement of a peculiarly fetid odor.

2. RÔLE OF BACTERIA IN THE NORMAL ORGANISM.

In a general way the microbes which live within or upon healthy individuals behave as commensals without giving rise to the slightest disturbance in their host.

Nevertheless, as there are almost always, among the germs distributed on the normal organism, a certain number which are pathogenic, we can understand that their presence constitutes a permanent source of danger for the economy. This danger exists upon all contaminated surfaces the germs being able to invade the organism when an accidental abrasion occurs.

Digestive action.—The germs lodged within the digestive canal have a more interesting rôle, for they seem to place themselves at the service of their host in order to assist in the preparation of his foods, in his digestion. Bienstock has isolated from fæces a bacillus which converts albumen into peptones; the bacillus amylobacter, which is found in the stomach of ruminants, secretes a soluble ferment which acts upon starch and cellulose.

M. Abelous has collected from the stomach sixteen species of microbes whose resistance to the action of the gastric juice he has verified. The study of their digestive action gave very interesting results. Albumen, casein, fibrin and gluten were rapidly and completely peptonized by several of them. Cane sugar was inverted by eight different species. Eleven species more or less completely converted starch into sugar. The conjoined action of these different micro-organisms upon a complex food must be considerable.

Putrefaction of dead bodies.—The healthy intestinal mucosa forms an effective barrier to the invasion of germs pullulating within the intestine; after death the cells having lost their power of resistance are rapidly dissolved by the diastases which these microbes secrete, and the latter penetrate within the tissues. They are first found in the peritoneum and on the surface of the abdominal viscera; they multiply in the blood of the mesenteric veins and extend along the portal vein, from which they progress toward the heart; thus, in various ways, they more or less rapidly invade the whole economy.

These anaërobic germs find in the organism deprived of oxygenated blood the most favorable conditions for their multiplication. Hence putrefaction is the more rapid in proportion as the blood is poorer in oxygen at the time of death, for example, in animals dead from charbon. In some cases the lack of oxygen is seconded by the absence of coagulation of the blood and of cadaveric rigidity, conditions which, by maintaining the fluidity of the medium, render microbic invasion more easy.

Putrefaction of cadavers is, therefore, primarily the effect of anaërobic germs coming from the intestinal surface; the anaërobes of other surfaces are in fact paralyzed in most cases by contact with the oxygen of the atmosphere. Putrefaction differs somewhat according to the surroundings in which it occurs.

Putrefaction in the air.—The multiple fermentations developed by microbes cause softening of the parenchyma, dissolution of the blood globules, infiltration of coloring matters in dependent parts, gaseous disengagements which permeate the connective tissue and elevate the skin, bloating, swelling of various parts of the body, rupture of the surfaces and oozing of sanious fluids. We then find aërobic germs implanting themselves in these fluids, consuming by the aid of the oxygen of the air the products of anaërobic fermentations, and thus completing the mineralization of the organic substance.

Such are the phenomena which supervene in bodies left in the air; they reduce these bodies to pulp in a few weeks, more or less rapidly according to the temperature.

Putrefaction under water.—In water, anaërobes only come into play; in running waters the surface of the cadaver is constantly being washed and superficial microbic invasion thereby prevented. On account of the preservation of the elasticity of the skin by constant bathing the swelling is more intense and more uniform. The accumulation of gases causes the body to float, and the blood accumulates in the dependent parts, in which alone cadaveric patches are produced.

Putrefaction in the soil.—Burial of a cadaver in a porous and absorbent soil is followed by absorption by the latter of the organic fluids as they are produced. From this results a comparative desiccation which maintains a certain degree of consistence in the body and interferes with microbic pullulation, whilst it favors the invasion of fungi. Hence, we see the development of molds (penicillium, aspergillus, etc.)

which excite a more complete combustion of organic matter; on the other hand, the dissemination in the soil of the fluid products of putrefaction renders the action of the anaërobic germs more general.

From a practical point of view these facts ought to be taken into account, as their rational application enables us to limit the intervention of microbes in putrefaction, several of which possess pathogenic properties and the emanations from which are in all cases to be avoided. The *earth-system* (burial of the cadaver in furnace-dried earth) realizes the ideal from this point of view.

Putrefaction is always slower in water and in the soil on account of the lower temperature of these media.

3. RÔLE IN THE ORGANISM IN THE PATHOLOGICAL CONDITION.

A certain number of microbes which meet with conditions favorable to their development within the organism of animals are the determining cause of diseases in these animals. These diseases are most frequently contagious. The pathogenic property of a microbe is a functional attribute of its vital faculties; in other words, it is to the life of the germ, to its nutritive requirements, to its secretions and excretions, and to its multiplication, that we must ascribe the disturbances which it determines. The study of pathogenic microbes, the principal object of this work, will be considered in the second and third parts.

VII. *Classification.*

Cohn has proposed to class the bacteria among the lower algæ. Previously they were generally considered as fungi. Several classifications of microbes have been proposed, none of which are perfect; we reproduce that of M. Guignard. The grouping of bacteria, however, has but little importance from our point of view. Nevertheless, it will assist the reader to a correct understanding of the meaning ascribed to certain terms in common use.

Arthrosporeae dividing	in one direction only				a) cells globular, in chains, surrounded by a gelatinous matrix	*Leuconostoc.*
					b) filaments, long, divided, without gelatinous matrix	*Beggiatoa.*
	in two directions				cells in distinct plates, adherent	*Merismopaedia.*
Endosporeae, dividing	in one direction only	a) cells globular	isolated or in chains, without gelatinous matrix			*Micrococcus*
			united		in a naked mass	*Punctula.*
					in a mass surrounded by a matrix	*Ascococcus.*
					in a mass forming a network	*Clathrocystis.*
		b) cells in the form of rods (bâtonnets)	isolated and short			*Bacterium.*
			united		in a naked mass	*Polybacteria.*
					in a mass surrounded by a matrix	*Ascobacteria.*
		c) cells in the form of small rods (baguettes)	straight		very short	*Bacillus.*
					filamentous, without sheath	*Leptothrix.*
					filamentous, with sheath	*Crenothrix.*
					branched	*Cladothrix.*
			spiral	free	very short	*Vibrio.*
					longer and rigid	*Spirillum.*
					very long and flexible	*Spirochæte.*
				united in a matrix		*Myconostoc.*
	in two or three directions					*Sarcina.*

PART SECOND.

GENERAL CONSIDERATIONS UPON PATHOGENIC MICROBES.

This secona part of our work will embrace: 1st. the study of pathogenic microbes in the static condition, that is to say, considered especially as to their general localizations, without taking into consideration their action upon the economy; 2d. the study of the reciprocal reaction of the organism and the microbe, which we will designate by the term physiology of pathogenic microbes; 3d. the transformations and destruction of microbes in their relations with hygiene and therapeutics, and 4th. finally, the methods of determination of pathogenic microbes.

CHAPTER I.

PATHOGENIC MICROBES IN THE STATIC CONDITION.

1. Conditions of life in external media and within the economy. Sources of infection.—2. Distribution of pathogenic germs.—3. Modes of contagion. Ways of penetration of pathogenic microbes.

Most micro-organisms live in the external world at the expense of dead matter; these have received the name of *saprogenic* or *saprophytic* germs.

Some live in either a permanent or adventitious way

upon animals and mankind; they are called *pathogenic* when, by their pullulation, they give rise to diseases. These affections have long been known but the study of their causes, the infinitely little, is of quite recent date; such diseases are called *infectious.* Later, we will have to distinguish infections properly so called from intoxications of microbic origin.

The term *virulence* is applied to the collection of properties by which pathogenic microbes are able to prove detrimental to living beings. This faculty is far from being unchangeable; it may present all degrees of intensity, in accordance with conditions capable of modifying the vitality of the germ.

The term *virus* is applied to solid, liquid, or gaseous vehicles containing pathogenic germs.

I. *Conditions of life of pathogenic germs in external media and within the living organism.*

Those germs which cause diseases do not exclusively and necessarily live within the economy. Some of them can also multiply in dead organic media; others, indeed, only pullulate within the organism incidentally the external media being really their natural field of multiplication.

In accordance with these considerations pathogenic germs have been divided into three categories:

a. Contagious obligatory parasites;

b. Contagious facultative parasites;

c. Non-contagious facultative parasites.

Contagious obligatory parasites.—These are represented by microbes which, under natural conditions, only reproduce themselves in the living organism and are transmitted from one animal to another without undergoing any evolution in the surrounding media. Of these, two varieties are distinguished; 1st. those which, little resistant, perish immediately on leaving the organism, and the transmission of which must be made directly by contact of a diseased subject with a healthy subject: syphilis, gonorrhœa, rabies; 2d. those which, more resistant, are preserved for a certain time outside of the economy, without however multiplying, and may arrive upon healthy subjects by means of various vehicles, contact with a diseased subject not being absolutely necessary: measles, variola, scarlatina, diphtheria, glanders, tuberculosis. In this case the source of the contagious disease resides principally but not exclusively in the diseased organism. If the germ is met with elsewhere it will be upon objects which have been in direct relation with the diseased or its cadaver; these objects can, moreover, communicate the latent germs of which they are bearers to the ordinary media: soil, water, or air, in which these germs preserve their contagiousness for a period of time more or less extended.

Whatever variations there may be in the mode of contagion, the contagious obligatory parasite arrives upon the healthy organism in the same condition in which it left the diseased subject where it had its origin.

Contagious facultative parasites.—These live and multiply not only within the organism of animals, but also outside of the latter, upon dead organic matters, in waters, etc.; germs of pyæmia, septicæmias, gangrene, erysipelas, typhoid fever, asiatic cholera.

The virulence of some of these germs seems even to be diminished by their passage through the organism to such a degree that the disease quickly dies out if their original virulence is not restored by a return to the outer media; it is thus with cholera, for example, which spontaneously disappears in winter in the countries of central Europe, probably because the external media have become unsuitable to its multiplication. In the case of cholera we perceive a gradation toward the miasms; but whilst in this last the microbe exhausts its effects upon the individual which harbors it and does not extend beyond it, the bacillus of cholera, on the contrary, proliferates during its passage through the organism of man and thus increases the chances of later infection.

The source of infection, therefore, for germs of this kind is twofold: the diseased organism and the infected media in which the germs pullulate; these two sources possess in the same degree the power of begetting the disease, and the opportunities of infection will be more frequent than for the germs of the first group in as much as the multiplication of the microbes in the different media is an important cause of their preservation.

Non-contagious facultative parasites.—These live, in the normal condition, in the external media, and it is only incidentally that they develop within the organism of animals. They do not seem in the latter to meet with conditions favorable to their vitality, for their effects are exhausted completely in the course of the disease which they determine, and the disease is not transmitted from one subject to another. The condition under which it occurs consists always in the impregnation of a healthy organism by germs drawn directly from external infected media; in short, the disease is bound to the soil: paludic fevers, perhaps yellow fever.

The three classes of germs which we have just been considering explain the distinction formerly made between the different forms of virus. The first corresponds to *contagions*, the second to *miasmatic-contagions*, and the third to the *miasms properly so called*.

Usually we also class with the miasms the toxic gases which are disengaged from cesspools, and the poison of the expired air.

Baumgarten has divided pathogenic microbes into *exogenous* and *endogenous*, according to whether they come from the exterior or from the diseased subjects themselves.

II. *Distribution of pathogenic germs.*

A certain number of pathogenic germs are ubiquitous in nature; we meet with them almost everywhere. Such is the case with the germs of suppuration and of septicæmias, on account of their excessive production and the readiness with which they live in the surrounding media. Specific germs are only incidentally found in such media.

In the following lines we will review in succession the various media which may contain germs noxious for the economy, this study being essential to a correct understanding of the different methods of contagion.

Air.—We have seen, in the first part, that the air can hold germs in suspension; it is natural to think that among these we may find some which are possessed of pathogenic properties. Moreover, it can readily be understood that the dust of the streets and accumulations of dirt will yield to the atmosphere, under the influence of numerous agitations of the air, particles contaminated with pathogenic germs, which may have been brought there in various ways.

It was for a long time believed that the air expired by the diseased was a fertile cause of the diffusion of pathogenic germs in the atmosphere, but we now know that the expired air never contains germs, neither have we succeeded in transmitting diseases in this way.

Transmission by the air has been accomplished experimentally for a certain number of diseases (sheep-pox, charbon, tuberculosis, vaccinia, etc.) by the dissemination of the dried virus of these diseases in the atmosphere.

Under natural conditions the atmospheric germs occur in a state of great dilution and hence the chances of infection are extremely limited. If we add that the air lends itself little or not at all to their multiplication, and that they are destroyed more or less rapidly by light, oxygen, and dessication, we shall see that the danger from the free atmosphere is almost nil.

The confined atmosphere of inhabited houses contains more germs than the external air, that of cities more than that of the country.

The confined atmosphere of infected places can become the carrier of disease germs and can transmit certain contagious diseases either by transporting these germs into the respiratory passages, or by depositing them upon alimentary matters or on the surface of wounds. The pyogenic microbes have been met with in surgical wards, the bacillus of Koch in those of the tuberculous. These germs come from the dried exudates, dressing cloths, etc.

It is therefore necessary, in places where diseased subjects reside, to take precautions against the germs of the atmosphere.

The condition of desiccation of virulent material has necessarily a great influence upon the richness of the air in pathogenic products, and all the conditions which assist in raising the dust are of such a nature as to increase the number of these products. Klebs has observed that during an epidemic of diphtheria a large number of new cases occurred after the sweeping of the streets and that they especially prevailed along the roads followed by the wagons used in transporting the dirt.

Waters.—From a pathological point of view, generally speaking, a water is the more to be feared the greater the proportion of organic matter it contains, since this material implies the presence of microbes which live at its expense. Most of the germs of water, however, are inoffensive; only rarely have pathogenic microbes been met with: septic vibrio, pus cocci, etc.

Pathogenic germs find entrance to waters in various ways; they may come from the air, from the bodies of animals which have succumbed to infectious diseases, or they may come from the soils traversed by the waters. This last mode has especially attracted attention during recent times: the bacillus of typhoid fever of man which is voided with the fæcal matters, can pass with the liquids of cesspools through very porous soils and thus come to contaminate the subterranean waters. When these waters are used for human consumption an epidemic of typhoid fever may result. The knowledge of this fact has enabled us in numerous cases to trace the disease to its source and check its extension. The bacillus of cholera is disseminated in the same manner.

Generally speaking, therefore, waters charged with organic matters should be viewed with suspicion, and it is necessary to take particular precautions in order to avoid the pollution of alimentary waters by cesspools, dung-hills, trenches of liquid manure, by the stagnant waters of the streets, etc.

The vicinity of these reservoirs of organic detritus to the sources of alimentary waters is always to be dreaded, especially when the earth is little adapted to filtration; hence, it is preferable, especially in thickly populated centers, to secure water which has filtered through virgin soils and convey it to the cities by a system of closed canals.

Pathogenic bacteria preserve their virulence in water for varying periods. It has been observed that this virulence is preserved during four months for the bacillus of charbon, one year for its spores; the bacillus of typhoid fever retains its vitality for two months in water, that of cholera for twenty-four hours in cesspools, and twenty-nine days in spring water; those of glanders and tuberculosis for twenty and ten days respectively (in water).

Soil.—The germs of the soil are numerous, their function being to transform organic matters of which this medium is the great recipient; but most of them are without effect on the organism of animals. Nevertheless, inoculation of vegetable mold into the small animals of the laboratory often rapidly leads to death with suppuration or gangrene as local lesions. Wounds contaminated by the soil readily become complicated by accidents of the same kind.

The specific germs which are incidentally met with in the soil are those of tetanus, charbon, Pasteur's septicæmia, typhoid fever, cholera, etc.

Germs are especially abundant in the superficial layers of the soil. The almost constant humidity of the soil in winter carries them into the deep layers; on the other hand, during periods of drought, in the absence of descending currents of water, the superficial washing is less complete and the germs remain more in the layers in contact with the atmosphere. Thus, other things being equal, the dust will be richer in germs in summer than in winter.

The micro-organisms of the soil become harmful in various ways: 1st. by contaminating the vegetable foods which grow upon an infected place; 2d. by distributing themselves in the air through the desiccation of the soil; 3d. by directly contaminating a solution of continuity (tetanus); 4th. by contaminating the waters which filter through the earth and which, later, are to serve for alimentation.

In the first three cases the germs must occupy the surface of the arable layer; microbes which are more deeply situated may return to the surface by the increase in height of the sheet of subterranean water, by the phenomenon of capillarity which constantly occurs in finely divided soils, or, finally, through the intermediation of earth worms, larvæ, etc., a fact which has been demonstrated by M. Pasteur in the case of charbon.

The infection of subterranean waters by the soil is subject to numerous influences dependent upon the location and nature of the land. Changes of elevation of the surface waters also produce effects which it is of importance to consider. It has been noticed that large floods are often followed by typhoid epidemics, a circumstance which has been attributed to the simultaneous rise in level of the waters of infiltration generally, and especially adjoining large sheets of surface water; this elevation of the subterranean waters brings them into contact with soils impregnated with putrid matters in the neighborhood of cesspools, trenches of liquid manure, sewers, etc., and when these putrid matters contain the typhoid germ there results a general pollution of springs and wells into which these contaminated waters diffuse themselves.

The pathogenic germs of the soil can retain their vitality for a period more or less extended; Grancher has seen the typhoid bacillus retain its vitality in the soil for five months. The charbon bacillus is also preserved in this medium, more especially, however, in the form of spores.

The destruction of these germs is dependent upon the action of the oxygen and light; as the conditions of their pullulation are more delicate than those of saprogenic microbes, the presence of the latter must also be taken into consideration in so far as their active proliferation more or less rapidly brings about an insufficient supply of nutrition for the pathogenic species.

Foods.—The vegetable foods may be contaminated by pathogenic germs either by contact with the soil while yet in growth (charbon, actinomyces), or, after harvesting, in the storehouses of fodders by the air of infected stables or barns, or from direct contact with sick animals.

MM. Galtier and Violet attribute to the fodder the development of the typhoid affections of the horse.

Foods of animal origin are sometimes bearers of pathogenic germs. This is the case when they come from subjects suffering from diseases caused by these germs. The flesh and all the tissues of animals dead from charbon contain the bacteridium.

Dwellings and vehicles.—Places which have been occupied by animals affected with contagious diseases, and vehicles (wagons, etc.) which have been used in their transportation are most frequently infected by specific germs. These are deposited upon the floor, walls and ceilings of houses and upon the mangers, racks, etc., either by the air or directly with the secretions or excretions of the sick. The transmission of microbic diseases through the intermediation of dwellings is therefore in every way possible.

Various articles and utensils.—All the articles and utensils which are found in places occupied by diseased animals or which have been employed in the service of the latter—harness, blankets, grooming instruments, sponges, brushes, curry-combs, etc., litter and manure—may become bearers of pathogenic germs which have been derived from these animals.

Healthy organism.—The bodies of animals in health are always bearers of microbes; but the latter are most frequently of no importance from our present point of view. We already know that the ubiquitous pathogenic germs—those of suppuration and of septic accidents—which are met with almost every-where—must be present on the surface of the body and even in its interior, upon the various mucous membranes and especially that of the digestive canal. Recently the bacillus of tetanus has been discovered in the fæcal matters of healthy horses. The pneumococcus—the pathogenic germ of lobar pnuemonia in man—is constantly present in the buccal mucus of nearly all individuals. The skin, the respiratory mucous membrane and that of the genito-urinary passages near the openings communicating with the exterior, are likewise contaminated with ubiquitous germs.

The transmission of a contagious disease through the intermediation of healthy individuals, animals or mankind, shows that pathogenic micro-organisms may be present on or in the body without producing disease. The methods by which infection is produced and the phenomenon of immunity sufficiently explain this peculiarity.

Infected organism.—The organism attacked by a contagious disease harbors the microbes of the latter in very different points according to the nature and localization of that disease. It is desirable, for the ends of a rational prophylaxy, to know the places of election of the germs of the different diseases, and, more especially, the seat of those which, in being eliminated from the economy, are able to contaminate healthy individuals.

We may meet with these germs in the various secretions flowing to the exterior: in the saliva: rabies; in the fæcal matters: tuberculosis, typhoid fever, cholera, chicken cholera, pneumo-enteritis; in the expectorations: tuberculosis, glanders, actinomycosis, sheep pox; in the urine: bacterial hæmoglobinuria of cattle, tuberculosis, etc.; in the semen and vaginal mucus: gonorrhœa; in the secretions of wounds: glanders-farcy ulcers, syphilitic chancres, lesions of dourine, pustules of variola of the different species of animals.

Hemorrhages may occur through the various passages and thus distribute externally the germs which are present in the blood: charbon.

If germs do not appear to be eliminated by the intact skin it is none the less true that the latter is frequently soiled by pathogenic germs emitted by the diseased; these germs are conveyed to the skin through contact of the secretions or infected litters.

A subject affected with a contagious disease does not appear to be at all stages of the disease capable of communicating the contagion to the same degree. Thus, in the case of glanders and tuberculosis, the danger appears to be absent if the softened lesions do not directly communicate with the exterior.

Cadavers of infected subjects.—The bodies of animals dead of infectious diseases are fertile sources of pathogenic germs; from the stand-point of the alimentary hygiene of man it is of great importance to recognize the place of election of those germs, but the question can hardly be considered in a general manner.

As to the duration of the virulence of pathogenic microbes in dead bodies left to themselves, it varies greatly according to the germs concerned, but upon this subject our knowledge is very incomplete. We know, however, that in some diseases the pathogenic power may persist for a long time—for years in the case of charbon and tuberculosis, for example. It is therefore necessary to carefully destroy these bodies.

III. *Modes of contagion.—Ways of penetration of pathogenic microbes.*

Those germs which are exciters of disease, after having multiplied in the bodies of the first infected animals, may be transported to other animals, and thus propagate the disease. This transference of the germs of a disease from a sick to a healthy individual constitutes contagion.

1. *Modes of contagion.* A microbic disease will the more surely be communicated, the contagion will have greater opportunity of taking effect, in proportion to the number of the germs emitted by the diseased. A superficially situated disease will be more readily transmissible, other things being equal, than a deep seated one; a disease with lesions of large extent will be more easy to communicate than one in which these are less extensive. The greater or less duration of the resistance of the virus to the natural agents of destruction is also a condition upon which depends the augmentation or diminution of the contagion-begetting power of the virus.

The transference of pathogenic germs from a diseased subject to a healthy subject—contagion—may be *direct* or *immediate*, that is, the healthy subject obtains the germs of the disease from the diseased animal itself, or it may be *indirect* or *mediate*, the healthy individual receiving the microbes eliminated by the diseased, through the intervention of external media.

Direct or immediate contagion.—Of this we have to distinguish two cases according as the transmission takes place between two subjects independent of each other (direct contact), or from the mother to the fœtus (heredity).

Direct contact.—Direct contact seems to be necessary for a certain number of diseases: rabies (bites), syphilis, dourine (coition), in which the germs, contagious obligatory parasities, appear not to maintain their virulence in the surrounding media. On the contrary, in other diseases direct contact is not essential, the microbes maintaining their virulence outside of the organism: glanders, tuberculosis, charbon, etc.

Man becomes infected by direct contact when he accidentally inoculates himself in making autopsy on bodies affected with glanders, tuberculosis or charbon.

Heredity.—Contagious diseases appear to be transmitted from the mother to the fœtus only by passage of the specific microbes through the placenta. The intervention of the father in phenomena of this kind is therefore indirect, in as much as the disease with which he is affected must first be transmitted to the mother, a fact which is observed in syphilis, for example. Here, therefore, the action of the mother in reality alone comes into play.

Some infectious diseases are transmitted from the mother to the fœtus; these are especially general affections and those in which the germs are able to circulate in the blood: certain septicæmias, charbon, fowl-cholera, strangles, rouget, tuberculosis.

The transmission seems to occur through alterations of the placenta, such as hemorrhages, specific lesions (tubercles, etc.), alterations which, indeed, are readily produced on account of the diseased condition of the mother and the tendency of the various germs living in the blood to determine vascular and other lesions.

Nevertheless, this interpretation, which certain observations sufficiently justify, does not explain all the facts of heredity, and it seems that a simpler mechanism intervenes in some cases.

We know that the contagious affections of the genital apparatus of the female are readily transmissible, and they will actually be transmitted to the fœtus unless their existence renders gestation impossible, or provokes abortion.

The disease transmitted to the fœtus may cause the death of the latter and its premature expulsion, it may disappear, or, finally, may remain in a stationary condition, permitting the complete development of the young animal, in which, at a later period, it may undergo fresh evolution.

Hereditary transmission may be limited to the communication of immunity by the diffusion through the placenta of soluble vaccine substances elaborated within the body of the mother. But immunity in the young being may also be consecutive to recovery from the disease with which it has itself been affected.

Indirect or mediate contagion.—We call the contagion indirect when the virulent germs which come from diseased subjects are transported on to healthy individuals after having been deposited on its surroundings. We have already seen in considering the distribution of pathogenic germs in external media, that they may be encountered in the air, water, the soil, foods, the walls of houses, on mangers, racks, manure, litter, and the various objects and utensils which have been more or less directly in contact with the diseased.

Now, these are just the infected vehicles which transport the pathogenic germs to healthy individuals.

Indirect contagion only occurs with facultative parasitic germs and with contagious obligatory parasitic germs which are endowed with considerable power of resistance against external causes of destruction.

The point of entry of microbes mediately transmitted is variable; when the virulent matters are brought into contact with the skin, or with the genito-urinary or ocular mucous membranes, it is called *transmission by indirect contact;* when the contagion is introduced with the food, water, or by the air, it is called *transmission by ingestion* or *inhalation.*

When pathogenic germs are once deposited on or within the organism, the manner in which they may affect the latter will differ according to the case; sometimes they will have no appreciable effect; at other times they may determine the irruption of a disease similar to that which has engendered them. The placing in activity of the pathogenic faculty depends in reality upon many circumstances, and, in the first place, upon absorption. We will study here the ways and processes of absorption of pathogenic microbes, as a sequel to the study of modes of contagion.

2. *Absorption of pathogenic microbes.*—Disease germs are capable of penetration through various surfaces, natural or artificial; these we will now review in succession.

Skin.—The intact skin is an unfavorable surface for the absorption of germs, but does not oppose itself in an absolute manner to their penetration; some of them are probably able to introduce themselves into its substance through the pilo-sebaceous glands, and thus give rise to diseases. The anatomical pustules which develop upon the hand or arm of anatomists, surgeons, and accoucheurs, seem to be produced in this way; it is the same with acne.

Repeated frictions of the skin at the time of contact with virulent matters will considerably further penetration. Garré induced the formation of furunculous pustules on his own arm by rubbing it with a culture of staphylococcus pyogenes aureus.

The incorporation of the virus with a fatty body, by rendering the contact more complete, increases the facility and certainty of absorption by the intact skin. Charbon and glanders have been communicated in this way.

In short, if absorption by the healthy skin is possible, it seems to take place only within narrow limits. It acquires, on the other hand, great importance when a traumatism has opened the way to microbes, most of these being capable of penetrating through even the slightest solutions of continuity in the external integument.

Digestive mucous membrane.—The different structure of this membrane in the various parts of the digestive canal manifestly implies variation in its absorptive faculty toward pathogenic microbes. The lining membrane of the anterior passages is little adapted to their penetration, but this much more readily occurs if one or several accidental solutions of continuity exist in these passages. Experiment has shown that the addition to fodders contaminated with the virus of charbon, of bodies capable of excoriating the mucosa (thistles, husks of barley) increases the mortality from this disease, but excoriations of the bucco-pharyngeal mucosa are comparatively common, and consequently it is necessary to take into account this way of infection.

Absorption, however, appears occasionally to take place in this situation in spite of the integrity of the mucosa, as, for example, when we contaminate a healthy flock with aphthous stomatitis by depositing a little of the saliva coming from diseased animals in the mouth of other healthy animals.

As in the case of the anterior, the mucous membrane of the posterior digestive canal admits of penetration by virus when it is the seat of alterations in its continuity: erosions, ulcerations. But, even when it is intact, the gastro-intestinal mucous membrane does not prevent the absorption of pathogenic germs, as numerous experiments have demonstrated. Charbon, glanders, tuberculosis, aphthous fever, cholera of fowls, and, indeed, nearly all microbic diseases can be transmitted in this way. Nevertheless, all subjects which ingest virulent products do not necessarily become infected; a certain number are refractory to the disease; in others the gastric juice kills all the germs which are introduced, whilst in others, again, absorption may not occur.

Referring to the microbicidal action of the gastric juice, we may repeat, that non-sporulated bacteria are killed much more rapidly than spores; infection by the latter is consequently infinitely more certain; this fact has been demonstrated for charbon.

Respiratory mucous membrane.—Air charged with

virulent dust, being inhaled by a healthy subject, may determine the outbreak of the disease, this fact having been established for tuberculosis, charbon, glanders, fowl cholera, etc. The experimental injection of microbes within the trachea gives the same result.

The respiratory mucous membrane, therefore, allows itself to be traversed by microbes. Nevertheless, the presence of pathogenic germs in the air does not imply that infection will necessarily take place. These germs are generally deposited upon the mucus of the anterior passages and are then rejected with the products of expectoration. The fact that germs are constantly absent from the expired air whilst they are always contained in the inspired air, sufficiently shows that the latter is purified in contact with the mucous membrane.

Absorption by the respiratory mucous membrane may occur throughout its whole extent; as with other lining membranes, this absorption is favored by solutions of continuity.

Ocular mucous membrane.—The conjunctival mucous membrane absorbs certain microbes; of this, accidental inoculation of gonorrhœal pus in the eye furnishes sufficient evidence. M. Galtier has succeeded in transmitting rabies in this manner.

Genito-urinary mucous membrane.—Syphilis, gonorrhœa, in man, and dourine in the horse are generally inoculated by coition in the absence of solutions of continuity either of the vagina or urethra. Frictions probably aid in the penetration of the specific germs. Absorption by these intact passages can, therefore, not be doubted.

The penetration of microbes, therefore, takes place

through the intact mucous membranes, but this faculty may give rise to no injurious effects upon individuals who are in absolutely physiological conditions. M. Bouchard has given the following explanation of the mechanism by which the normal organism opposes itself to the penetration of the germs present on the natural surfaces:—Germs which have traversed the epithelium immediately come into conflict with the white corpuscles distributed in the derm of the mucosa, and by these they are taken up and digested. If, on the contrary, the economy is disturbed, thrown out of equilibrium, by a sufficient cause (repeated influence of currents of cold air, influence of fear) the enfeebled white corpuscles allow the microbes to break through the barrier which they are charged with defending. The experiments of Bouchard were made with ordinary germs; their results are applicable to pathogenic microbes, with this difference, that it is necessary, here, to take into account a supplementary factor directly related to the pathogenic faculty: we refer to the influence that the secretions of these microbes can themselves exert, in such cases, upon the white corpuscles, in attenuating or annihilating their action.

Wounds.—In general, wounds are the surfaces of predilection for the absorption of pathogenic germs not only on account of the division of the tissues and vessels which, in a manner, opens the way for them, but also because the organism is not prepared with its means of defense. The latter has to be organized upon the field, and but too frequently proves inadequate to repel the invaders. Nevertheless, hemorrhage, and the phagocytic action of the elements of the tissues, as well as that of the leucocytes which speedily accumulate in the wound, are conditions unfavorable to absorption. The latter, moreover, depends upon many other circumstances bearing upon the nature of the germ, its vehicle, the depth and extent of the wound, etc. The microbes of tetanus and those of gaseous gangrene, for example, only multiply in wounds to which the access of the air is limited; their activity is checked by atmospheric oxygen. The rapidity of penetration is influenced by the nature of the medium; an aqueous medium will be more readily absorbed than a solid excipient or one of thick, colloid consistence.

Tuberculosis, symptomatic charbon, and grangrenous septicæmia are not inoculable by sub-epidermic punctures, whilst this inoculation is successfully performed in the subcutaneous cellular tissue. Absorption is always more easy when the tissue is itself lacerated.

Absorption from wounds is, in general, very rapid; glanders has been seen to supervene in spite of deep cauterization of the inoculated wound two hours after the insertion of the virus; cauterization after a lapse of ten minutes has still allowed the evolution of sheep-pox. The amputation of the ear of a rabbit inoculated with charbon by sub-epidermic puncture in that region has not prevented the irruption of the disease, although this operation followed only three minutes after the inoculation.

Infection by wounds may be local only, or it may become generalized; in the latter case the extension occurs chiefly by the lymphatics, the germs then showing their presence in these vessels by the lesions which they determine in the corresponding glands. But pathogenic germs can also penetrate directly into the blood-vessels in which case generalization occurs much more rapidly.

CHAPTER II.

PHYSIOLOGY OF PATHOGENIC MICROBES.

1. Action of pathogenic microbes upon the organism. Receptivity. Immunity.—2. Reaction of the organism against pathogenic microbes. Phagocytosis. Bactericidal condition.—3. Evolution of the bacterial disease.

I. *Action of microbes upon the organism.*

MECHANISM OF THE PATHOGENIC ACTION.

Pathogenic germs exert their action upon the economy in two principal ways, of themselves, or by their secretion products. We will consider, successively, the mode of development of the troubles which they occasion, both local and general.

1. *Pathogeny of local alterations.*—In local lesions the microbe acts at first like a foreign body, that is, it excites a purely mechanical irritation; the Koch bacillus very probably acts in this way when it gives rise to tubercular neoplasms; we know, indeed, that the injection of lycopodium powder into the blood develops a similar lesion, and the egg of the *strongylus vasorum* appears to act in the same way in the pseudo-tubercles which it occasions in the dog.

This mechanical influence, however, seems to be much the least important. The soluble substances—diastases and ptomaines—secreted by the microbes possess very active properties and to these we ascribe the genesis of most of the manifestations of the infectious diseases.

Microbic diastases, as we have already said, represent the digestive juices of the microbes; they can therefore act chemically upon the substance of the tissues, provoking hydrations and chemical decompositions, and thus lead to the liquefaction, softening, and even mortification of these tissues. These dissolutions are sometimes preceded by a sort of coagulation (coagulation necrosis).

The soluble substances—diastases and ptomaines—may possess phlogogenic properties; in this case they develop, of themselves alone, all the inflammatory phenomena which follow the inoculation of the microbes from which they come. They excite swelling, proliferation, and fatty, hyaline, or waxy degeneration of the elements, dilatation of the vessels, exudation of products more or less plastic, sometimes active diapedesis of the white blood corpuscles and even extravasations. The pneumobacillus liquefaciens bovis, isolated by Arloing from the lesions of bovine contagious pleuro-pneumonia, secretes a diastase which excites inflammatory œdemas recalling those of the disease itself. The staphylococcus pyogenes aureus produces a diastase and a ptomaine which are phlogogenic and pyogenic.

The local alterations which supervene in consequence of the introduction of certain germs into the tissues sometimes depend, at least in part, on the chemical action to which the nutrition of these germs gives rise; the emphysematous tumors of symptomatic charbon and of traumatic gangrene are caused by the abnormal production of gas which accompanies the fermentations provoked by the anaërobic microbes of these diseases.

2. *Pathogeny of remote and general manifestations.*—Anatomical alterations remote from the original point of infection may be produced as a consequence of the penetration of the pathogenic microbes into the lymphatics and blood-vessels; their pathogeny is identical with that of the local manifestations.

The arrest of the microbes in vessels of small caliber may become the starting point of secondary mechanical lesions; we refer to microbic embolisms, comparatively frequent in general infectious diseases, and which are followed by infarcts, stases, etc. But it may happen that the secretion-products alone penetrate into the circulation, the bacteria remaining intrenched at the primary focus; we may also have localized alterations dependent upon special properties, dissolving, phlogogenic, or pyogenic, of these products.

Besides the functional troubles resulting directly from the anatomical lesions, primary or secondary, seated in the various organs, we have to consider the genesis of the general manifestations which accompany microbic diseases, either local or general.

Fever is one of the most frequent symptoms; a certain number of soluble microbic substances excite *hyperthermia.* This almost always results from a general nutritive excitation of the tissue elements, from contact with these substances; but it may also be the result of a diminution in the loss of heat through constriction of the peripheral capillaries. Rise of temperature therefore does not, in all cases, imply increased production of heat.

Several microbic substances are known to possess this pyretogenic or fever-begetting property: those derived from the bacillus of blue pus, from Freidlander's microbe, etc.

Further, the soluble products of the anatomical elements themselves produce similar effects. According to Gangolphe and Courmont, necrobiosis of the tissues develops pyretogenic substances independent of all microbic intervention. These authors have observed that *bistournage* is followed by fever when the testicular products are able to penetrate into the blood, whilst fever is absent if the precaution be taken to put a ligature completely around the scrotum so as to prevent all absorption. It is not impossible that the super-activity of the phagocytes, struggling against microbic invasion, becomes also the starting point of the production of pyretogenetic substances, a hypothesis which certain facts seem to indicate. We know, moreover, that extracts of flesh and of the spleen possess similar properties, which thus appear to belong to a certain number of the residues of normal disassimilation.

Some soluble substances of microbic origin excite the phenomenon of *hypothermia*, lowering the temperature of the body; this property is possessed by the soluble products of the comma bacillus of cholera, the septic vibrio, and the staphylococcus aureus.

The nervous system is especially sensitive to the action of these products of microbic origin, in the way of excitation (hyperkinesia developed by the toxines of tetanus), or of depression (coma, somnolence, in fowl cholera and charbon; paralysis consecutive to diphtheria and to the pyocyanic disease). The heart, the respiratory centers, and the vaso-motor center may also feel the influence of these substances, from which may result sometimes an increase, sometimes a diminution, of the functional activity of these organs.

RECEPTIVITY.

Receptivity, or aptitude to contract infectious diseases, varies in accordance with a large number of circumstances, and, especially, with the species and the mode of inoculation, with the individual, age, heredity, causes of depression, the quantity and the quality of the virus, and the association of the virus of more than one disease.

1. *Influence of species and of the mode of inoculation.*—The susceptibility of a given animal species to a disease, when experimentally inoculated, does not necessarily imply the liability of that species to contract the disease spontaneously.

Tuberculosis develops spontaneously, with great ease and frequency, in man, cattle, and birds; it is much rarer in the horse, the pig, and the dog under the same conditions; as for the rabbit and the guinea pig, which it is very easy to inoculate experimentally, they do not contract the disease except by inoculation.

Symptomatic charbon, appearing spontaneously only in bovines, is inoculable to the sheep, the goat, and the guinea pig, but not to other animals.

Rabies, a spontaneous* disease of the dog, is inoculable to all species of mammals and to birds, intracranial inoculation being always successful, whilst hypodermic inoculation gives only variable results.

Bacteridian charbon develops naturally in cattle, sheep, and horses, experimentally in all the domesticated mammals.

Influence of the individual.—All the individuals of any species susceptible of contracting a microbic disease do not take that disease when they are exposed to the contagion; some of them are less susceptible than others and a few may even be absolutely refractory.

Influence of age.—Strangles, and distemper of dogs are diseases of youth; symptomatic charbon only appears in cattle of from six months to four years and is with difficulty inoculated to the young calf; guinea pigs are the more susceptible to charbon the younger they are. Pigs of less than four months are much less susceptible to rouget than adults; hence, only very young pigs should be vaccinated against this disease.

4. *Influence of heredity.*—Certain predispositions are inherited; for example, that of children born of consumptive parents, to contract tuberculosis.

5. *Influence of depressive causes.*—All conditions which have a debilitating effect on the organism facilitate microbic invasion. Pasteur has shown that lowering the temperature of fowls inoculated with charbon is followed by the evolution of the disease in these animals in spite of their natural immunity.

* [Acquired by transmission from its own kind.—D.]

Bouchard has shown that gradual cooling of the guinea pig allows the entry into the blood of this animal of the microbes distributed on the surface of the mucosa. Nervous disturbances of a depressive character act in the same way: fear.

The administration of chloral, curara, or alcohol to the frog and to the dog, which are refractory to bacteridian charbon, endows them with receptivity for this disease. Antipyrine and chloral diminish the resistance of the chicken to the same affection.

The attenuated and inoffensive bacillus of symptomatic charbon becomes pathogenic and causes the disease when it is conjoined with lactic acid, or trimethylamine, or when injected into a tissue previously contused. The same substances, as well as a similar traumatism, also favor the implantation of the tetanus bacillus.

The influence of these very diverse conditions must be ascribed to the action which they exert upon the phagocytes, the protective function of which they diminish, by rendering them less capable of digesting the microbic invaders.

6. *Influence of the quality and quantity of the virus.*—Receptivity naturally varies with the special degree of virulence of the germs that the organism receives; a germ which is toxic for a given species may become an efficacious vaccine when its virulence has been enfeebled.

As to the quantity of the virulent substance, it especially requires attention; the influence of the dose can not be questioned; the natural immunity of Algerian sheep against charbon is overcome by inoculation of a large dose of the charbon bacillus. Inoculation of a minimum quantity of certain germs confers immunity, whilst a larger dose produces the fatal disease (gangrenous septicæmia, symptomatic charbon).

7. *Influence of microbic associations.*—The susceptibility to an experimental disease is increased by injecting, at the beginning of this disease, a considerable amount of the soluble products coming from the microbe inoculated (Bouchard). This fact has been established for the charbon bacteridium, fowl cholera, the staphyloccus aureus, bacillus prodigiosus, the bacillus of symptomatic charbon, the pyocyanic bacillus, etc. This is the more striking since the injection of the same products without microbes often confers immunity. Here, on the other hand, it aggravates the trouble or renders it possible, by overcoming the natural or acquired immunity. The mode of action of these substances consists, in this case, in the obstruction which they oppose, by paralyzing the vaso-dilator nerves, to the diapedesis and phagocytosis which the germs of the disease naturally excite when they are inoculated to vaccinated subjects, or those which are naturally refractory.

The adjuvant action that the microbic secretions exert in association with the microbe from which they come, may also be manifested in association with other germs; thus, the natural immunity of the rabbit against symptomatic charbon is obliterated if we inject to this animal, at the same time with the charbon bacteria, a sterilized culture of the staphylococcus aureus, or micrococcus prodigiosus.

The same interpretation is applicable to the predisposing or aggravating influence that a previous or concomitant disease exerts upon the course of another infectious disease.

In the case of tetanus, however, the association of accessory germs, such as the bacillus prodigiosus, renders more certain the irruption of the disease, by provoking diapedesis, because the medium then becomes more favorable to the multiplication and nutrition of the tetanus bacillus.

On the other hand, the presence of phlogogenic germs, which excite an excessive diapedesis, may oppose the receptivity for a given microbe. Thus, recovery from malignant pustule nearly always occurs after the appearance of suppuration ; this is explained by the fact that the white corpuscles, whose accumulation is brought about by the foreign germs, overcome the bacteridium.

Receptivity of the tissues and organs.—Pathogenic germs do not act in the same manner upon all parts of a susceptible organism ; they appear to have a sort of affinity for certain organs, for certain tissues. Thus, the bacillus of bacteridian charbon multiplies in the blood whilst that of symptomatic charbon only develops in the connective and muscular tissue and is, on the contrary, killed in the blood.

The pneumococcus generally limits its field of operations to the lung; the bacillus of pneumoenteritis to the lung and the intestine; that of glanders to the respiratory apparatus and to the skin, etc.

This elective action of pathogenic microbes has, in some cases only, received a scientific explanation.

When the germs have penetrated into the blood they will localize themselves preferably in the organs predisposed, enfeebled, hence less effectively prepared for defense.

IMMUNITY.

Immunity consists in the absence of receptivity; it may be *natural* or *acquired*. Acquired immunity is most frequently consecutive to a first attack of the disease. A large number of microbic affections do not recur; they leave behind them organic changes, inappreciable directly, but more or less permanent, which oppose themselves to the later development of their germs; the individual, once recovered, is vaccinated. This is the case in variola, vaccinia, sheep pox, syphilis, charbon, rouget, fowl cholera, pleuropneumonia, etc. Other diseases, on the contrary, can reappear several times in the same individual; such are gonorrhœa, simple chancre, diphtheria, erysipelas, tuberculosis, glanders, etc. In the case of these two last diseases we see, in reality, specific lesions supervene at periods very remote from the time of the first appearance of the affection, which would not happen if one or more first attacks of the disease had vaccinated the organism. Immunity may also be communicated by artificial means, and the process by which this is effected is known as vaccination.

The duration of acquired immunity varies greatly according to the disease and the individual concerned. In general, immunity consecutive to the natural disease is more permanent than that conferred by vaccination.

Patient researches have thrown a little light upon the nature of immunity, and shown that it operates, at least in part, through the active intervention of the organism which possesses it.

It was at first supposed that the non-recurrence of infectious diseases was due to the fact that the microbes, at the time of the first attack, had abstracted from the blood principles indispensable to their growth, or, on the other hand, contaminated the blood with principles which opposed their growth. In short, two theories were entertained: that of exhaustion and that of impregnation.

The first is to-day abandoned; immunity has, in fact, been overcome by employing large doses of virus, which would be impossible if the organism was really impoverished in substances indispensable to the microbes.

The doctrine of impregnation, on the other hand, is strongly supported by the discovery of vaccinating substances. A certain number of microbes secrete substances which mix with the fluids of the tissues, diffuse through all the economy, and impair the vitality even of the germs which have produced them; these *preventive* substances which appear not to be identical with the toxic secretions of microbes oppose themselves to a recurrence of the disease.

The existence of soluble vaccinating substances has been unquestionably established for a certain number of diseases: the blue pus disease, bacteridian charbon, symptomatic charbon, cholera, Pasteur's septicæmia, pneumo-enteritis of the pig, etc.

Impregnation of the blood does not account for the persistence of immunity. Substances prejudicial to the growth of the germs must gradually become eliminated from the economy, and if vaccination were due solely to their presence, its effects would be extinguished in a comparatively short time.

The doctrine of the impregnation of the humors is happily supplemented by that of the modification of the solid parts of the organism, the anatomical elements.

In acquired immunity, whether it be consecutive to a first attack of the disease or be conferred by vaccination, the fluids and tissues possess bactericidal properties. Thus, the serum of the rabbit vaccinated against the pyocyanic disease, is bactericidal for the microbe of this disease; but this property also belongs to the solid tissues of the vaccinated animals, a fact which has been further demonstrated by the failure of attempts to cultivate the bacillus of symptomatic charbon in the thigh of a guinea pig vaccinated against this affection. The production of the bactericidal state by vaccination is established for a certain number of germs: the bacillus of bacteridian charbon, symptomatic charbon, blue pus, the vibrio of cholera, and Metschnikoff's vibrio.

This property is communicated to the vaccinated animal by the mingling of the vaccinating substances with the nutritive liquids of the economy; the contact of these fluids with the tissue elements brings about a permanent nutritive modification of the latter; this modification, throughout all the time of the duration of the immunity, exerts its influence upon the fluids of the body, endowing them with the microbicidal faculty. By virtue of this faculty the virulent germs against which the vaccination has been directed, when they attempt to invade the organism, find themselves opposed, attenuated, through contact with the fluids, their secretion products which tend to diminish diapedesis are less abundantly produced and the phagocytes, as a result of the reaction which naturally occurs on contact with foreign bodies, issue from the vessels and take advantage of the situation.

The bactericidal state, therefore, enfeebles the virulent microbes, which are then overcome and removed by the phagocytes.

Immunity being in some way fixed in the anatomical elements, one can understand its persistence and the possibility of its hereditary transmission. In natural immunity or the refractory condition the bactericidal property does not appear to have, at least in certain cases, the same importance as in acquired immunity. Indeed, the blood of an individual refractory to a given microbe can serve perfectly for the artificial culture of this microbe. Nevertheless, too much importance should not be attached to this fact. We shall see, later, that the blood is always more or less bactericidal, but that this faculty disappears shortly after its removal from the vessels. The bactericidal property of the blood, therefore, depends, without doubt, upon its contact with the living tissues, hence we can form no accurate estimate of this condition of the blood contained in the vessels of the organism by comparing it with the same fluid outside of the vessels.

We know, again, that natural immunity results, in the case of certain microbes, on account of the temperature of the organism being either too high or too low; it may also, as M. Arloing has demonstrated, be the result of a natural insusceptibility of the organism to the action of the amorphous products secreted by the microbes.

II. *Reaction of the organism against microbes.*

We will consider, successively, under this head: phagocytosis, the bactericidal property, the isolation and elimination of pathogenic microbes, and variations of virulence produced by the organism.

When non-pathogenic germs, or germs in which the virulence is extinguished, are introduced into the tissues of an animal, these germs are more or less quickly destroyed. The same thing occurs when pathogenic germs in full virulence are introduced into the system of an individual destitute of receptivity for the germs. On the other hand, if the virulent germs are brought into contact with an organism endowed with receptivity they will multiply and become the starting point of morbid troubles.

There are, therefore, in the organism of certain animals, conditions capable of bringing about the destruction of microbes. These conditions are multiple and their nature is as yet incompletely elucidated. We know, however, that the organism defends itself against invasion through the intermediation of its figured elements and of its fluid parts.

1. *Phagocytosis.*—The name *phagocytosis* is given to the destructive action of certain cells toward microbes, these cells being known as phagocytes. The white corpuscles, in this regard, take first place in whatever part of the body we find them: blood, connective tissue, lymphoid organs (spleen, lymph glands, marrow of bones, etc.); then come the fixed cells of the connective tissue, the endothelial cells of the capillary vessels, the cells of soft epithelia, muscular fibers, etc.

The disease develops when the phagocytes do not succeed in killing the introduced germs; in the opposite case it fails to develop.

The injection of virulent germs in a susceptible individual paralyzes the phagocytes; on the contrary, the injection of attenuated germs is followed by an accumulation of leucocytes (diapedesis) around the place where the germs occur; the germs are then taken into the substance of the leucocytes and digested by them. The same thing happens when we inject virulent germs into a non-susceptible subject.

All the germs seized by the phagocytes are not infallibly destroyed; hence, these migratory cells, in certain cases, seem to be a means of transferring bacteria from one part of the organism to another.

The leucocytes possess, in common with some of the lower vegetables, the property (called *chemotaxis*) of moving themselves toward certain chemical substances. Now, experiments have shown that the sterilized or filtered cultures of a certain number of microbes attract the white corpuscles, whilst cultures of other microbes either have no action upon white corpuscles or paralyze their movements.

Some germs secrete substances which paralyze the vaso-dilator nerves, thus opposing a direct obstacle to diapedesis and therefore to phagocytosis. As an illustration of this nervous action we find that the ear of the rabbit which has received an injection of the soluble products of the pyocyanic bacillus, for example, does not become inflamed when a layer of croton oil is applied to it.

Microbes whose secretions attract the leucocytes * will be more easily overcome by them, and will be less dangerous for the economy.

Microbes which repel the white corpuscles, finding themselves in good conditions as to nutrition and surroundings, will break down the barrier which the corpuscles oppose to them. But, under conditions unfavorable to their development, their secretions diminish and the phagocytes regain all their power.

Some agents, such as chloral and chloroform, are capable of suspending the chemotaxic faculty of the phagocytes.

Certain influences—physical and moral disturbances, fatigue, nervous perturbation, cold, which often cause the irruption of an infectious disease or aggravate it—have a depressing effect upon the vaso-dilator nervous apparatus, interfere with diapedesis, with phagocytosis, and therefore favor the implantation or multiplication of the germs of disease.

2. *Bactericidal or microbicidal property.*—By the bactericidal property is meant the peculiar quality of the humors of the economy—blood, aqueous humor, pericardial serosity, etc.—which impedes or prevents the multiplication of pathogenic bacteria in these fluids. This bactericidal faculty varies greatly according to the species, the individual, and the germs with which we have to do. When microbes are introduced into the blood, a certain number of them perish; those

* [According to Buchner, the attractive action (positive chemotaxis) exerted by sterilized cultures of certain microbes toward leucocytes is dependent on the proteid contents of the bacterial cells, rather than on their secretion products. Baumgarten's Jahresbericht, 1890.—D.]

which resist are then capable of multiplying; hence, the normal microbicidal property of the blood is temporary, not permanent. Fresh blood kills the bacillus of charbon, but constitutes a suitable medium for its cultivation eight days after its removal from the vessels. It seems to be established that, when we inject these bacilli into the blood, they multiply there only after having previously multiplied in one or more organs in which they have been arrested. When only a small number of microbes have been introduced they may all be destroyed, and then the inoculation fails.

Non-pathogenic microbes, introduced into the blood, disappear from this fluid, becoming arrested in the fine capillaries of the liver, spleen, marrow of bones, and kidneys, in which situation they are quickly destroyed.

Pathogenic microbes have a similar experience when they are inoculated in small doses into the circulation; like the preceding, they are consigned to the fine capillaries of the parenchymatous organs and there sustain the conflict with the phagocytic elements (endothelial cells, white corpuscles, etc.) According as the issue of this conflict proves favorable to the microbes or to the phagocytes we may expect the appearance or non-appearance of the disease.

The evolution of the microbic disease is accompanied, when the disease is non-recurrent, by the production of the bactericidal property. This condition is slowly evolved by the progressive action of the soluble microbic products upon the fluids and tissues. When it has acquired sufficient intensity it may lead to the limitation of, or recovery from, the disease, by opposing itself to the multiplication, and interfering with the nutrition of the introduced germs; the phagocytes then assume the duty of freeing the organism from these enfeebled germs.

Elimination of microbes.—The infected organism may free itself from microbes by the local reaction which their presence excites. The germs of pus, for example, determine around them the diapedesis of leucocytes and these destroy by phagocytosis a large number of germs. On the other hand, at the same time that the accumulation of leucocytes within the meshes of the vascular network presents a certain degree of obstruction to the nutrition of the microbes, the microbic secretions act upon the leucocytes, either by virtue of their dissolving disastases or of their toxic products, so that a parallel destruction ensues of white corpuscles and of microbes. Whilst this double destruction goes on at the center of the focus, the surviving germs continue to invade the peripheral tissues until there is formed all around the invading army a sufficient barrier of phagocytes. We then see the limitation of the abscess, and, occasionally, its encystment by fibrous organization of the limiting tissues; but more frequently the extension of the pyogenic inflammation gives rise to softening, perforation of the integument, and the elimination of the pus and therefore of the germs, the cause of all the trouble.

The natural elimination of pathogenic microbes can occur in all cases in which the lesions they determine are situated near the surface of the skin or mucous membranes communicating directly with the exterior; this elimination is more or less complete according to the case.

When the germs are deeply situated their expulsion may still take place through certain glands—the kidneys, the salivary glands (rabies), the liver (symptomatic charbon). The elimination of infectious germs by way of the kidneys has been observed for a large number of diseases; generally, if not always, it is the consequence of the irritation which the microbes or their products exert upon the organ. Many microbic diseases, in fact, are accompanied by nephritis. The more or less complete desquamation of the secretory epithelium, and the vascular troubles which then supervene account for the passage of the microbes.

If the elimination of the figured agents of the infectious diseases demands conditions somewhat complex, the rejection of their secretions is more readily accomplished, since these substances are dissolved in the liquid media of the economy. Their filtration takes place especially through the kidneys; their presence in the urine has been established; but they may also transude from the various other natural emunctories.

4. *Modifications of virulence*—The bactericidal property of the humors has the power of attenuating the virulence of pathogenic microbes. Inversely, the absence of this property may occasion an increase of this virulence. Thus, the virulence of a given microbe may become increased or diminished by passing through a series of individuals of the same species: rouget of the pig becomes more active when it is made to pass through the pigeon or the rabbit;

on the contrary, the organism of the ape attenuates the virus of rabies.

The virulence, modified by one species, may be changed in the same or opposite direction for other species: rouget which has become more virulent in the rabbit is less virulent for the pig; whilst the passage of the same germ through the pigeon augments its pathogenic power not only for the pigeon but also for the pig.

We could multiply examples of the influence of the natural organic media upon pathogenic germs. The virus of rabies attenuated in the ape is also attenuated for the dog, guinea pig, and rabbit. Inoculated in series to one of these species, it regains the virulence which it naturally possesses in the dog; in the guinea pig and rabbit it may even surpass this, and in the latter attain a maximum of activity beyond which further passages no more modify it.

III. *Evolution of the bacterial disease.*

After what we have seen of the reciprocal action of the organism and pathogenic microbes, little remains to be said of the disease itself.

The determining cause of microbic affections resides always in the implantation of the specific germ in a susceptible individual; but such common causes as cold, mental emotions, overwork, etc., may take an important part in their etiology, by diminishing or suspending the normal phagocytic action, as we have already seen.

The germs once introduced and having resisted the combined influence of phagocytosis and the bactericidal property, their effects do not immediately become appreciable. A preparatory period then ensues during which the microbe proliferates, multiplying its means of action upon the economy. This period corresponds to the *incubation*. It is shorter the better the organism is adapted to the life of the germ; its duration also varies with the nature of the germ and its virulence, with the abundance of the virus, with the receptivity of the subject, the place of inoculation, etc. In some diseases the duration of the incubatory period is almost constant; in others it is very variable (rabies).

Certain diseases have several successive incubations, or, rather, latent periods, during which the disease germ ceases its activity, slumbers, to resume at a later period its course of invasion (tuberculosis, dourine, syphilis). The organism is then in the condition of *latent microbism.*

The period of incubation is completed when the first manifestations of the morbific action of the virus appear; the premonitary symptoms are in no way characteristic; yet, in the course of an epizootic, there are certain signs by which we can recognize the *invasion* of the disease in an individual previously healthy. Thus, a persistent fever in one of the cattle of a stable where pleuro-pneumonia prevails, would excite a suspicion of its invasion by this disease.

The early obscure symptoms more or less quickly give place to troubles more and more serious, which clearly characterize the affection with which we have to do, and express the influence of the progressive action of the microbe. This is the period of *increase.* We will not enter here into details of the common symptoms of microbic diseases, an outline of which has been incidentally given in connection with the subject of their pathogeny. We will only say a few words on the *specificity* of pathogenic microbes. At an early period in bacteriological research investigators applied themselves to the discovery of a special microbe for each disease; it was believed that every germ produced always the same effects. We now know, however, that the same microbe may give rise to very different diseases: the microbe of fowl cholera gives a true septicæmia in the rabbit, a circumscribed abscess in the guinea pig; the streptococcus pyogenes develops sometimes an abscess, sometimes an erysipelas, sometimes puerperal fever. Similarly, a given lesion may be consecutive to the inoculation of different microbes: the particular inflammatory lesion which has received the name of tubercle has for its usual cause the tubercle bacillus; but other agents can develop identical changes, which have been designated pseudo-tubercles in order to distinguish them from those in which the said bacillus exists; among the number of these agents we may mention croton-oil, lycopodium spores, the demodex folliculorum, strongylus vasorum, the utricular sarcospermia of the muscles of the pig, the actinomyces, the pseudo-tubercle bacillus of Courmont, etc. Typhoid fever of the horse appears to develop under the influence of various species of microbes, acting independently.

Microbic specificity is, therefore, not absolute; it depends upon the organism in which the parasite has implanted itself and on the external conditions which may have influenced the latter. Attenuation of a microbe suffices, indeed, to change its effects.

An infectious disease may be local or general, according as the germs are confined to the place of inoculation or have invaded the circulation. In the latter case the general affection may have been preceded by circumscribed local lesions, or it may have been generalized from the first. In either case, a general disease can determine localized lesions, specific or not: nephritis, hepatitis, enteritis, inflammatory enlargements, etc. These secondary inflammations of the secretory organs may be the starting point of grave complications (*auto-intoxications*): absorption of bile, urinary intoxication. Changes of these organs present, in addition, direct obstacles to the elimination of the soluble microbic products.

The multiplication of disease germs arrives, at the end of a certain time, at its apogee; the disease then reaches its height. The secretion-products of the microbes become harmful to the microbes themselves; in mingling with the fluids they communicate to them, as well as to the tissues, the faculty of checking microbic proliferation, the microbicidal faculty; in a word, they vaccinate the organism. The vaccinating effect is slowly produced under the influence of the prolonged contact of these products. Impoverishment of the organic media in principles indispensable to the microbes, febrile elevation of temperature, can also act prejudicially upon the latter. From this ensemble of unfavorable circumstances there results enfeeblement of the germs, and the phagocytes take upon themselves the task of destroying them.

The disease then subsides on account of the fact that the toxic substances of microbic origin which become eliminated in various ways, are no longer replaced by fresh additions. Recovery is not complete when the elimination of these substances is ended. Besides the weakness of the patient, local troubles more or less important may persist, recovery from which will take place gradually now that the cause which engendered them is removed.

But the disease, having attained its height, may terminate in death, the manner in which this is brought about varying in different cases.

Microbic diseases may be *acute* or *chronic*, *epizootic*, *enzootic* or *sporadic*. The gravity of some, at least, of these diseases is subject to variations, whilst others of them are almost invariably fatal: charbon and rabies for example. Indeed, from the knowledge which we have gained in regard to receptivity, immunity, and the resistance of the organism we should expect all degrees of intensity in such diseases. In epizootics of great severity it is observed that the first animals attacked are more severely affected than the last, and that the number of the individuals attacked, considerable at the beginning, rapidly diminishes toward the end of the attack. This is probably due to the fact that the virus fixes itself upon individual animals by reason of their special susceptibility. The least refractory will be first attacked, will be severely affected, and the virus will rapidly multiply, thus increasing the chances of infection for all those which are susceptible to the disease. On account of this dissemination of the virulent germs the less susceptible will also be finally stricken, but, by reason of the fact that they do not offer a very favorable field, the microbic pullulation will in these be less extensive, the disease less serious, and the chances of infection for the individuals yet exempt and which are, moreover, the most refractory, will rapidly diminish.

The disease may be *continuous* (charbon), *remittent* (tuberculosis), or *intermittent* (malaria).

CHAPTER III.

TRANSFORMATION AND DESTRUCTION OF PATHOGENIC MICROBES IN THEIR RELATION WITH HYGIENE AND THERAPEUTICS.

1. Morphological and physiological variations of pathogenic microbes.—2. Attenuation.—3. Preventive inoculations. Vaccinations.—4. Destruction and annihilation of pathogenic germs.

1. *Morphological and physiological variations of pathogenic microbes.*

We have already, in the first part of this work, referred briefly to the influence of the media upon microbes in general. This influence acquires great importance when pathogenic germs are in question. The latter may undergo considerable changes under the influence of external conditions.

Pathogenic germs undergo, from this cause, variations of form and of function.

Physical variations are quite common: the microbe of avian cholera shrinks when its cultures become old; that of rouget, cultivated after its passage through the rabbit, increases in volume; the bacillus of the pyocyanic disease presents itself, according to the media in which it is cultivated, as a bacillus, a spirillum, or a micrococcus; the bacillus of gangrenous septicæmia grows in the form of short rods in the connective tissue, in long filaments in the serous membranes and blood. The bacteridium of charbon, cultivated in bouillon containing a small quantity of bichromate of potassium, loses its power of forming spores, subsequent generations to which it gives birth being also asporogenous. A temperature of 42° to 43° produces the same effect upon the bacteridium.

The functional variations are of more importance; the pyocyanic bacillus, placed under certain conditions, ceases to secrete the coloring matter which characterizes it. There are, however, variations of virulence which more especially interest us. The virulence of pathogenic germs may be increased or diminished, then brought back to its normal intensity, by conditions which vary with each microbe; these conditions will be considered more in detail in the following paragraph.

The virulence of a microbe may become enfeebled to such an extent as to completely lose its pathogenic powers; it then becomes saprogenic. From the recognition of this fact to the admission of microbic transformation there is only a step. Nevertheless, up to the present, we have not observed the formation of a new species at the expense of another species. From a practical point of view, however, it must be admitted that certain saprogenic species may incidentally become pathogenic and determine the development of morbid troubles in man or animals.

II. *Attenuation of pathogenic microbes.*

The diminution of virulence of pathogenic microbes is occasionally, not always, connected with a diminution in their general nutritive activity; it may occur under quite varied conditions, either spontaneously or from certain definite influences purposely brought into play by the experimenter; we will here study the different means by which attenuation may be obtained.

1. *Attenuation by the normal atmosphere.*—We already know that the majority of the atmospheric germs are dead. It is logical to admit that the loss of their vitality did not take place abruptly, but, on the contrary, was gradually produced, and that what virulence they may originally have possessed also disappeared gradually. In short, the germs of the air are attenuated before being destroyed. The atmospheric conditions which determine these changes in microbic life are far from being simple: oxygen, light, the electrical condition, desiccation, probably all act in concert.

But the external air, of itself alone, can bring about the attenuation of pathogenic germs. Very active cultures may lose their virulence in some days, weeks, or occasionally months.

Cultures of fowl cholera, abandoned to the air, gradually diminish in virulence so as to completely lose it at the end of a time varying from six weeks to two months, occasionally much less. By re-sowing these germs, in way of attenuation, at periods more and more remote from the establishment of the culture, we obtain generations of progressively decreasing virulence, a series of viruses less and less powerful, the special activity of which is preserved when we exclude them from contact with the air or when we rejuvenate them without intermission by cultures made at very short intervals. Attenuation is here, therefore, hereditary through successive generations. Those of most feeble virulence constitute vaccines against more virulent cultures.

Cultures of the germ of rouget undergo changes similar to those of fowl cholera.

As to cultures of the charbon bacillus, they are infinitely more resisting to the destructive influence of the atmosphere. This is because they contain spores, and when we wish to attenuate them in contact with the air it is necessary to begin by preventing sporulation. Pasteur has attained this end by cultivating charbon at a temperature of 42° to 43°. At this temperature, the culture exposed to contact with the air, rapidly loses its virulence ; it ceases to be fatal first for the larger animals, then for small adults, finally, for small animals only a few days old.. The bacteridium itself perishes much more slowly. Now, each degree of virulence can be perpetuated by cultivating at 42° to 43° the different varieties obtained, each of these varieties transmitting its special virulence to its descendants. If they are returned to 37° they form spores possessing in embryo the special pathogenic activity of the attenuated bacilli from which they originate, and susceptible of transmitting the latter to new generations cultivated at 37°.

According to M. Chauveau, the attenuation of the

bacteridia by the preceding process is the effect of the heat and not of the oxygen, as was advanced by Pasteur.

The least virulent varieties produce a mild disease which leaves behind it immunity for the varieties less attenuated, and we thus have vaccines of different degrees of strength which we can bring into action in succession.

The attenuated virus of fowl cholera and of bacteridian charbon can be restored to their normal virulence by passing them through the organism, first, of young individuals, and then through those of individuals of gradually increasing age.

2. *Attenuation by compressed oxygen or air.*—M. Chauveau has succeeded in attenuating the bacteridium of charbon by subjecting it to contact with pure oxygen at a pressure of two and a half atmospheres during fifteen days and over, at a temperature of 35° to 36°. This attenuation is transmitted through successive generations of the bacteridia. He has, in this way obtained an energetic vaccine which confers immunity against charbon without giving rise to accidents in the vaccinated, the vaccinating property being wholly retained whilst the virulence becomes progressively enfeebled until it entirely disappears.

3. *Attenuation by heat.*—Heat is a powerful means of destruction of germs; we also find it among the number of attenuating agents. It is to Toussaint that the honor belongs of having first demonstrated this property of heat in connection with charbonous blood. His methods were improved, and Chauveau applied himself to the task of heating small quanti-

ties of charbonous blood in such a manner that all the bacilli were influenced to the same extent; the degree of enfeeblement, in fact, depends on the temperature and the duration of the heating. Chauveau has prepared two vaccines, intended to be inoculated in succession: the first, the least active, is obtained by heating defibrinated blood at 50° during fifteen minutes; the second, by heating the same liquid at the same temperature during nine to ten minutes. This last vaccine inoculated to animals which have not received the first may yet cause the death of some individuals.

To obtain a uniform result with charbonous blood it is necessary always to use fresh blood in which the bacilli are free from spores. These latter are endowed with a much greater power of resistance and are not attenuated like the bacilli; hence, in his researches upon the attenuation of cultures, M. Chauveau first cultivated the bacteridia at a temperature of 42° to 43°, in order to prevent the formation of spores. The asporogenous bacteridia are heated to 47° during three consecutive hours; they are then attenuated to the extent that they no longer kill adult guinea pigs. Brought then to the favorable vegetating temperature of 37° the attenuated bacteridia form spores; but these are liable when placed under suitable conditions to produce virulent bacilli again.

In order to render attenuation transmissible to successive generations of bacteridia M. Chauveau then brought the sporulated cultures to temperatures neighboring on 80°. The attenuation then became fixed upon the spores and the latter, returned into nutritive bouillon at 37°, produces bacilli attenuated like themselves. Of two vaccines prepared in this way, the first comes from heating at 84° during one hour, the second from heating at 82° during the same time.

The natural serosity of the specific lesions of symptomatic charbon may also be attenuated to different degrees by a temperature of 65° to 70°, maintained for a greater or less period of time (Arloing, Cornevin and Thomas). However, these authors work in preference with serosity dried at a temperature of 30° to 35°; the dried virus is, in fact, more fixed than the fluid serosity; it can be preserved indefinitely with its normal virulence; its attenuation requires temperatures varying between 60° and 110°. The authors have prepared two vaccines from it: one heated at 100°, the other at 85°; Kitt has recommended a single vaccine heated at 90°.

4. *Attenuation by solar light.*—Light is a powerful bactericidal agent; by the careful use of this property M. Arloing has succeeded in gradually attenuating the charbon bacillus.

"Thus, a culture exposed to the rays of the sun during nineteen hours furnishes a virus which kills the guinea pig in the dose of one drop; exposed during twenty hours, a culture only kills one guinea pig out of two; exposed during twenty-five hours, such a culture no more kills guinea pigs but vaccinates them, the vegetating power of the bacillus being also considerably diminished."

The diminution of virulence thus obtained is temporary and is not transmitted to later generations of the bacilli.

5. *Desiccation.*—Desiccation has been utilized by M.

Pasteur in attenuating the virus of rabies. The cords of rabid rabbits, suspended in vessels containing pieces of caustic potash and maintained at a temperature of 20°, lose their virulence in seven days; the diminution takes place progressively, so that, by fixing the successive degrees of virulence by inoculation to the rabbit, we can obtain virus of gradually increasing intensity.

Attenuation is here the result of several factors: desiccation, oxygen, and the temperature. The influence of the last agent is itself very great if we can judge by the increased rapidity with which attenuation is produced when the temperature is even slightly increased; the virulence is, in fact, obliterated in seven days at 20°, in five days at 23°, and in 24 hours at 35°.

6. *Attenuation by antiseptics.*—Chemical substances which are toxic for bacteria can diminish the vitality of the latter when they are employed in selected doses and their contact maintained during definite periods of time.

The bacillus of symptomatic charbon is attenuated to the point of becoming a vaccine in contact with sublimate, at 1 to 5000, with carbolated glycerine, eucalyptol, and thymol. Carbolic acid, at 1 to 800, allows the multiplication of the charbon bacteridium but prevents sporulation. Continuation of the contact with this solution gradually weakens the virulence. The same results are obtained, but much more rapidly, with bichromate of potassium in the proportion of from 1 in 2000 to 1 in 5000.

The soluble substances secreted by germs can diminish the virulence of other germs; thus bouillon containing the residues of a culture of cholera attenuates the charbon bacillus.

7. *Attenuation by passage through the organism of animals.*—Pathogenic germs are subjected, by the organisms upon which they implant themselves, to certain influences of which we have already had occasion to speak; we have noted especially the modifications which may supervene in the virulence of these germs either in the way of increase or diminution.

Instances of attenuation being produced under these conditions are quite numerous.

The microbe of rouget of the pig becomes well acclimated in the pigeon and the rabbit, and in these two species acquires great virulence; but, whilst the organism of the pigeon renders it more active for the pig, that of the rabbit diminishes its virulence for this animal. This attenuation is preserved in cultures then made in ordinary bouillon, and these cultures can be employed as vaccine for the pig.

The bacilli of bacteridian charbon and of symptomatic charbon are attenuated by their passage within the lymphatic sacs of the frog.

The virus of rabies is attenuated in passing through the organism of the ape to the extent of becoming inoffensive for the dog and of vaccinating it. It results from the preceding facts that the morbigenic faculty of microbes can be lessened in different degrees; in certain cases this diminution only exists for those bacteria upon which the conditions determining the attenuation have acted, and the attenuation is temporary; in others the modification to which the microbes have been subjected is more profound, more durable, and persists in later generations of these microbes.

The methods which furnish an hereditary attenuation permit of obtaining more easily large quantities of attenuated virus, and are more especially utilized in the preparation of vaccines.

III. *Preventive inoculations. Vaccinations.*

The organism may be made refractory to a bacterial disease by different means; by inoculating the natural virus, the attenuated virus, a chemical vaccine, or, finally, a virus different from that against which it is desired to fortify the organism.

1. *Preventive inoculation of natural virus.*—It has been observed that a certain number of contagious diseases leave behind them in the subject, after recovery, a solid immunity against these same diseases. On the other hand, certain diseases purposely communicated show themselves much less dangerous than when they prevail naturally. The recognition of these facts has given origin to variolization, clavelization,* preventive inoculation against pleuro-pneumonia, etc. The first, for more than a century, has been replaced by vaccination; the second is still recommended in our times, in default of a means of prevention more practical, if not more efficacious.

In this case the only object is to produce a disease of the same nature as the spontaneous disease, but mild in character, not threatening the life of the individual, yet endowing it with immunity. This end is obtained by *diminishing the number of the germs*

* [Artificial infection of flocks with ovine variola.—D.]

which engender the disease; the organism overcomes a small dose of a certain virus when it would be overcome by a larger dose: a minimum quantity of the virus of gangrenous septicæmia and of symptomatic charbon vaccinates against these diseases; a larger quantity produces the fatal disease. Dilution of the vaccine, of the virus of sheep-pox, lessens its effects.

The severity of a disease may, again, be diminished by *introducing the virus by a special way* known to mitigate its influence: the blood (pleuro-pneumonia, symptomatic charbon, gangrenous septicæmia, rabies), the cellular tissue of the tail (pleuro-pneumonia).

2. *Preventive inoculation of attenuated virus.*—Attenuated viruses develop a mild disease which confers on the animals an immunity more solid the less the degree of attenuation; in general, we have recourse in practice to several specimens of virus of different degrees of intensity; we begin with the weakest and end with the strongest; a solid immunity may thus be communicated by virus sometimes yet very active but against the action of which the less virulent varieties have fortified the organism.

Attenuated viruses which are able to prevent the development of infectious diseases are called *vaccine viruses* or simply *vaccines.*

The two charbons, rouget, pneumo-enteritis of the pig, chicken cholera, rabies, etc., have furnished vaccines the employment of which is to-day admitted into general practice.

3. *Preventive inoculation of soluble vaccinating substances.*—We have seen above, in connection with the subject of acquired immunity, that the latter results from the impregnation of the organism with the solu-

ble products produced by pathogenic bacteria. These vaccinating substances are known for a certain number of diseases—the pyocyanic disease, bacteridian charbon and symptomatic charbon, cholera, Pasteur's septicæmia, pneumo-enteritis of the pig, rabies, human tuberculosis (Koch's lymph will give immunity to guinea pigs), tuberculosis of birds (Courmont and Dor have vaccinated the rabbit by means of the soluble products of cultures), etc.

These substances, also called chemical vaccines, have a great advantage over the figured elements; the attenuated virus may indeed, in exceptional cases, regain its virulent properties, and this unknown to the experimenter; it then produces the fatal disease instead of the immunity which was expected of it. The soluble vaccinating substances, however, are not entirely free from danger; they are nearly always extremely violent poisons, and the quantities employed must be judiciously regulated. The discovery of these substances is of such recent date that in practice we have not as yet reaped the benefits from them which we may reasonably expect.

Attempts at the prevention and cure of the microbic diseases by organic liquids coming from species naturally refractory to these diseases have been made in recent times; here, again, the action concerned is a chemical one. The blood of the goat, transfused to the rabbit at the time that the latter is inoculated with tubercular products, will prevent the evolution of tuberculosis; if the transfusion is made after the disease has already commenced, it will retard it and may even cause its retrogression.

Injection of the blood serum of the dog (hæmocyne)

vaccinates the rabbit against tuberculosis; this preservative action is more intense with the serum of a tuberculized dog and manifests itself even when the injection is made seven days after the virulent inoculation.

We must mention here, in connection with preventives of a chemical nature, the attempts at vaccination against rabies with the essence of tansy, against tetanus with strychnine, and against tuberculosis with tannin.

4. *Preventive inoculation with the virus of another disease.*—Cow-pox is preservative against variola; we place the example in this paragraph although, according to recent researches, it should have its place along with the inoculations of attenuated virus. Recent experiments of M. Eternod have, in fact, demonstrated the identity of variola with vaccinia. It should be said, however, that these results have been contested.

The attenuated microbe of fowl cholera vaccinates fowls against charbon and against Davaine's septicæmia; guinea pigs vaccinated against symptomatic charbon are also vaccinated against the septic vibrio, but the reverse does not hold good.

IV. *Destruction and annihilation of pathogenic germs.*

The destruction of pathogenic germs is a very important point to consider. The hygienist and the physician should apply themselves to the suppression of dangerous germs wherever they exist, that is, upon the patient and upon the objects which have been contaminated in his surroundings; this operation is known as *disinfection*. Nature comes powerfully to their aid in this work of purification; but it often requires to be seconded by artificial means, the application of which generally abridges the natural duration of pandemic or panzootic diseases. We have therefore to separate the causes which bring about the destruction of pathogenic germs into natural and artificial.

It may be stated as a fundamental principle that adult bacteria in their vegetative form are more rapidly destroyed, whatever be the cause of destruction, than the spores or organs of fructification. For the vegetative forms the action of destructive agents is more or less rapid according to the case; a bacterium, taken in full vegetation in a suitable medium, will be more easily killed than one in the condition of latent life, for example, in a state of desiccation, which is equivalent to saying that death will be more easily produced as life is more intense, more complete. Something of the same kind is observed in higher beings, the liability of these to suffer from adverse conditions being greater as their requirements are the more exacting. A young bacterium recently developed from a spore is more sensitive than an adult bacterium which has attained its complete development.

1. *Natural disinfecting agents.*—These are light, desiccation, and oxygen.

Light.—Light excites oxidation of the organic substances which enter into the constitution of germs, and, as a consequence, involves the death of the latter. Other conditions being the same, solar light acts more rapidly upon non-sporulated germs than upon spores, upon germs contained in a liquid or moist medium than upon those which are in a dry medium.

The time necessary for destruction by light varies from a few hours to some weeks, according to the case; solar light, however, is a sure agent and one whose beneficial action operates in a continuous manner; hence, we should guard against voluntarily depriving ourselves of it; abundance of light in inhabited places is one of the most rational of hygienic measures.

Desiccation.—Insufficiency of water arrests the multiplication of microbes; the latter then lose their vitality more or less rapidly. But the spores resist much longer than the bacteria themselves; we know, indeed, that the virus of symptomatic charbon is dried in order to preserve it, and that the germs of tuberculosis are preserved active for a long time in pulverulent sputum.

Oxygen.—The germs of the air are destroyed more or less quickly under the combined influence of desiccation, which arrests their pullulation and impairs their vitality, of oxygen, which oxidizes them, and of the solar light, which excites in them this oxidation; the oxidizing action of oxygen, without doubt, extends itself to the bacteria of waters and of the soil, these perishing more or less rapidly by reason of the conditions unfavorable to their multiplication which they meet in these media.

Other natural agents, less important, may intervene to destroy pathogenic bacteria; among these we will notice again the ordinary saprogenic germs which, finding themselves in the same media as the pathogenic, may lead directly to the death of the latter; we have already explained this action.

2. *Artificial disinfecting agents.*—These are heat and the antiseptics.

Heat.—Heat is a powerful means of disinfection. Although germs withstand extremely low temperatures, they are very sensitive to increase of temperature. Temperatures too low, as well as those which are too high, are unfavorable to the proliferation of pathogenic microbes; but the latter are, in addition, rapidly fatal. Nevertheless, the spores are infinitely more resistant than the adult germs; whilst these last are killed in all cases between 50° and 100°, the spores, on the contrary, require a temperature of 110° to 125°. However, we can with certainty kill all spores in a given infusion by bringing, on different occasions, and at intervals of one day, the said infusion to a temperature fatal for the adult form, and taking the precaution between the periods of heating to place the whole at a temperature favorable for vegetation. The spores which the boiling has not been able to attack become transformed into adult bacteria which the heating of the next day will destroy.

Heat acts with more rapidity upon bacteria in the moist condition than when dried, upon those of water than upon those of the air.

Whether we have to do with adult germs or spores, with a dry medium or a moist, the duration of the heating is of great importance and can be supplementary to insufficiency of temperature.

Antiseptics.—This name has been given to chemical substances capable of destroying pathogenic germs or of arresting their development. The number of the antiseptics is quite considerable, but there are some which stand at the top of the list and merit more especially the confidence of the practitioner. These are, indeed, violent poisons for nearly all microbes, whilst the others, less energetic, are only really efficacious when they are directed against microbes the vitality of which is easily destroyed. Nevertheless, we can not establish a general rule upon this subject; the antiseptic action depends not only on the chemical agent employed, but also on the microbe concerned; it is also dependent upon the duration of contact, the nature of the excipient, etc. The alcoholic solutions of carbolic acid, for example, are less antiseptic than aqueous solutions of the same concentration because the former have a smaller power of penetration than the latter.

Antiseptic substances act either by rendering the media unsuitable for the multiplication of the microbes, or by opposing themselves to the production of the diastases indispensable to the elaboration of their foods.

We believe that it may be profitable to reproduce here the classification of the antiseptics proposed by M. Miquel. The antiseptic substances are here classified according to the dose of the substances necessary for the sterilization of a litre of beef bouillon:

1. *The most powerful antiseptics:* ·01 to 0·1 gram.
   (Oxygenated water, corrosive sublimate, nitrate of silver.)
2. *Very strong antiseptics:* 0·1 to 1 gram.
   (Iodine, bromine, sulphate of copper.)

3. *Strong antiseptics:* 1 to 5 grams.
   (Bichromate of potassium, chloroform, chloride of zinc, carbolic acid, permanganate of potassium, alum, tannin.)
4. *Moderate antiseptics:* 5 to 20 grams.
   (Arsenious acid, boric acid, chloral hydrate, salicylate of sodium, sulphate of iron.)
5. *Weak antiseptics:* 20 to 100 grams.
   (Borate of sodium, alcohol.)
6. *Very weak antiseptics:* 100 to 300 grams.
   (Arseniate of potassium, iodide of potassium, sea salt, glycerin.)

CHAPTER IV.

METHODS OF DETERMINATION OF PATHOGENIC MICROBES.

1. Examination, investigation, and staining of bacteria.—2. Culture of bacteria.—3. Experimental inoculations.

We have now, in order to finish the general study of pathogenic germs, to review the different methods of determination of these germs, that is, the means which we have of distinguishing them from each other, or recognizing them. This means consists in, 1st, microscopical examination of the bacteria either directly or after coloring them; 2d, their culture in artificial media; and, 3d, the test of the effects which they determine in animals.

In practice, and for certain germs which are well

characterized, the first of these means is often sufficient; but in other cases the germs to be determined must be subjected to the test of the three reactions which we have just mentioned. Thus, for example, the bacillus of human tuberculosis and that of avian tuberculosis possess identical physical characters, react in the same way to coloring matters, and give cultures difficult to differentiate; on the other hand, they have different effects on experimental animals. In default of this last test we might have concluded that the two germs were identical when they are really quite distinct in some of their physiological properties.

I. *Examination, investigation, and staining of bacteria.*

The first examination of a liquid, from a bacteriological point of view, should always be made without the help of reagents; we thus see the germs in their real form and size. Desiccation, or the addition of staining solutions often change the morphological character of microbes. It is in fresh preparations, also, that we are in a position to judge of the mobility or immobility of the beings which we wish to study.

Microbes being naturally colorless, little refringent, and consequently difficult to distinguish, staining methods have been adopted in order to bring them into relief in the media to be examined.

For this purpose we have recourse to the colors derived from coal— aniline colors—which have an intense staining power. Without entering further into the details of this subject, we may say that, with the stains employed, we often associate adjuvant substances which act the part of mordant, that is, which increase the penetrating power of the coloring substance and render its fixation more energetic; such are, for example, alum, potash, soda, lithia, carbonate of ammonia, aniline, etc., all of which we will find in the various liquids used in histology and microbiology.

Heat is calculated to expedite the staining of microbes, as well as of organic tissues in general; we shall see that this property has been taken advantage of in order to obtain rapid staining of various bacilli, and especially those of tuberculosis.

Two groups of aniline colors are recognized. In the first (*basic* colors) the coloring substance is associated with a colorless acid. In the second (*acid* colors) it plays the part of acid and is combined with a colorless base.

This distinction should be noted because the colors of the first class have a special affinity for cell nuclei and for bacteria, whilst those of the second give diffuse staining, that is, fix themselves indifferently on all parts of the cells and on the intercellular substance.

The basic colors include *gentian* and *methyl violet*, *methyl green*, *methylene blue*, *fuchsin*, *bismark brown*, *vesuvin*, and *safranin*.

The principal acid colors are *eosin* and *fluorescine*.

Single and double staining.—In any preparation containing microbes we can give to all the elements, microbic and non-microbic, the same coloration. In the case of certain micro-organisms we ought, indeed, to limit ourselves to this method of staining, such as the germs of fowl cholera, pneumo-enteritis of the pig, etc. With these germs we can, at most, only attenuate the color of the tissue elements with the aim of bringing more distinctly into view the deeper color of the bacteria. Such is the principle of a process of single staining which has been applied with slight variations by different micro-biologists, and which will be further referred to later.

Patient researches have placed at our command another method, called double staining. This method furnishes excellent preparations and renders great service in the study of the distribution of microbes in the tissues. Double coloration is obtained by different means, of which the choice is not indifferent, for the same microbe does not take the stain by all of these processes. The manner in which microbes behave toward the methods of double staining is even utilized for characterizing them.

The principle of double staining is as follows: 1st, the whole preparation is stained uniformly; 2d, all the elements other than bacteria are decolorized whilst the latter retain, on the contrary, the color which was originally imparted to them; 3d, the parts decolorized by the preceding operation are stained with a color which stands out distinctly from that of the microbe. Thus, the latter being violet, the contrast color will be red or brown; if, on the contrary, the microbes are stained red, the background of the preparation should be stained blue or green.

We may have to examine liquids, or solid tissues. In this last case it is sometimes necessary to make a previous examination of the organic pulp and study it later in thin sections. We shall therefore consider, in succession, the various manipulations to which the liquids, pulp, and sections of organs must be subjected.

Liquids—Direct examination.—A drop of the liquid is deposited upon a slide, covered with a cover glass, and at the margin of the latter is placed a drop of coloring matter; penetration gradually takes place, a process which may be assisted by manipulation of the cover glass. We may also add a trace of coloring matter directly to the drop of liquid to be examined before inclosing it under the cover glass. For this purpose aqueous solutions of gentian violet or methyl violet are generally employed.

Examination after desiccation.—The preparation just described has its field of usefulness, but it has the defect of not being permanent, and, especially, of not permitting of double staining. To attain these ends the liquids are rapidly dried so as to fix the elements which they contain and cause them to adhere to the glass. In this process a drop of liquid is spread on a cover glass either by means of a spatula or by pressing the drop between two cover glasses which are glided upon each other and then separated. The cover glasses, thus coated, are held over the flame of a spirit lamp or placed upon a hot plate, the surface containing the liquid to be examined being directed upward so as not to be attacked by the flame. The drying is sufficient when the liquid is transformed into an opalescent layer; this layer then intimately adheres to the cover glass, which can be then transported through a series of reagents without risk of the former becoming detached.

The quantity of liquid employed should be very

small in order that the dried preparations may form a very thin coating, which, after staining, will not obstruct the passage of the rays of light. This recommendation is of special importance in the case of organic fluids containing many histological elements, such as pus, for example, of which only a very small particle should be taken.

Before all staining, it is sometimes necessary to remove the fatty matters from the dried substance by immersing it in chloroform or a mixture of equal parts of alcohol and ether; this practice is indispensable in operating upon milk.

Preparations of dried blood are also much improved by this treatment: under the influence of the alcohol and ether mixture the corpuscles acquire a stability which desiccation can not confer upon them, and they then present very distinct forms in stained preparations, in which their relations with the microbes can be studied. This is a highly commendable practice which much increases the precision of researches bearing upon the blood. It should be followed by a second drying.

Organic pulps.—These are obtained by scraping the freshly cut surface of the organ suspected of containing microbes; a small particle is spread out between two cover glasses in a thin layer, in the manner indicated above, and then dried.

Microscopic sections.—Tissues which are intended to be cut for the study of the distribution of the microbes which may be present, should, first of all, be well fixed, that is, their elements well immobilized as regards their form and relations, and thus placed beyond the reach of cadaveric changes. This is especially necessary in the examination of those tissues for micro-organisms, as these latter, living independently, may continue to multiply after the death of the tissues, and, further, in organs left in contact with the air foreign germs generally develop which are prejudicial to the accuracy of such researches. These accidents are avoided by immersing the tissues to be preserved in absolute or, at least, strong alcohol, the precaution having been taken, in order to render the action of the alcohol more rapid and more intimate, to cut the organs into small cubes of half to one centimeter on the side.

Sectioning is performed in various ways, but the technique of this operation can not be described here; its study belongs to the domain of histology. Moreover, this mode of research does not appear to be within reach of all practitioners on account of the instruments and time which it requires, and, in nearly all cases, it can be omitted when the differential diagnosis of infectious diseases is the only object in view.

Mounting of preparations.

Cover glasses.—After staining, the cover glasses must be dehydrated to admit of their being mounted in Canada balsam. This end is attained very rapidly by drying them afresh over the flame of the spirit lamp or upon the hot plate, after having previously drained them between two folds of filter paper. As soon as the cover glass is dried, the preparation is cleared by depositing successively upon the coated surface a drop of cedar oil, clove oil, or bergamot oil, a drop of xylol, and, finally, a drop ot Canada balsam dissolved in xylol. The cover glass is then placed upon the slide in such a manner that the balsam distributes itself between the two glasses.

We can omit having recourse to clearing (clove oil, cedar oil, etc.), and, after drying, pass directly to the balsam; but it is incontestable that, if this practice increases a little the duration of the work, it also gives more distinctness to the preparation.

Sections.—Stained sections must also be dehydrated in order to be mounted in balsam. For this purpose they are passed successively through specimens of alcohol of different strength, ending with absolute alcohol (70, 90, 100 per cent). They are then cleared by immersing them in one of the substances mentioned above (cedar oil, bergamot oil, clove oil), then placed in xylol, and, finally, on a drop of balsam which has previously been deposited on a slide and which is now at once covered by the cover glass.

These diverse operations may be made upon the slide itself, the stained section having been spread and fixed at the start by a semi-desiccation. This last is easily accomplished either by the aid of filter paper which is used to absorb the water from the section, or by moderate heating.

Methods of single staining.

Hydro-alcoholic solutions of aniline colors.—These are made extemporaneously by adding a few drops of a saturated alcoholic solution of the dye to a watch glass of distilled water. They are largely used for single staining, especially for cover glasses. This method is sufficient for a rapid examination, when the object is merely to determine the presence or absence of microbes in the preparation.

The solutions most used are those of gentian violet, methyl violet, fuchsin, and methylene blue. A few minutes contact is sufficient: five for cover glasses, fifteen for sections, a little longer in the case of methylene blue.

Löffler's method.—Cover glasses are stained for from five to ten minutes in the following medium:

Solution of caustic potash, 1 to 10,000,	3	cub. cent.
Saturated alcoholic solution of methylene blue, ...	1	" "

All the elements are stained deep blue. To bring out the microbes, the stained preparations are immersed in water containing a small quantity of acetic acid, one or two drops to a watch glass; in this they remain on an average only one minute; they are then washed with distilled water and mounted. The microbes appear colored a deep blue, the tissue elements a light blue.

Method of Malassez and Vignal.—Staining is here made with Malassez' blue, which consists of:

Aniline water, ...	9	cub. cent.
Absolute alcohol, ...	1	" "
Saturated alcoholic solution of methylene blue, ...	1	" "

Sections remain therein ten minutes, cover glasses five minutes; they are then decolorized in the following mixture:

Absolute alcohol, ...	1	cub. cent.
2 per cent solution of sodium carbonate,	2	" "

Methods of double staining.

Gram's method.—Gram's violet contains:

Aniline water, ...	10	cub. cent.

Absolute alcohol, ... 1 cub. cent.
Saturated alcoholic solution of gentian violet, ... 1 " "

Cover glasses remain in this fluid five minutes, sections fifteen, after which they are transferred for one to two minutes into *Gram's solution of iodine in iodide of potassium* which consists of:

Iodine, ... 1 gram.
Iodide of potassium, ... 2 grams.
Distilled water, ... 300 grams.

They are then completely decolorized in absolute alcohol, and the ground-work of the preparation stained red (eosin, picro-carmine) or brown (bismark brown, etc.) The violet remains fixed on the microbes, but in the tissue elements it is replaced by the red or brown.

Weigert's method.—*Weigert's violet* consists of:

Saturated aqueous solution of methyl violet 6 B, ... 68 grams.
Absolute alcohol, ... 11 grams.
Aniline, ... 3 grams.

Sections are stained from five to ten minutes, immersed for one to two minutes in Gram's iodine solution, passed rapidly through absolute alcohol (a few seconds) in order to remove the greater part of the water, and decolorized in aniline oil; they are then cleared in xylol and mounted in balsam. Since by this method contact with alcohol is avoided, it is necessary to stain the groundwork of the preparation in red or brown before staining with violet. This inversion may also be employed without inconvenience in staining by the Gram method. The employment of alcohol in dehydrating may be entirely avoided in another way; to this end, the section, after coming from the iodine, or even before staining, is spread out, moist, upon the glass and fixed there in a state of semi-desiccation by slightly heating it or by the aid of tissue paper. On the section, thus fixed, are then deposited in succession a drop of the different reagents, the excess of each being carefully removed with the tissue paper before replacing it with the next.

In the methods of Gram and Weigert as well as in the method by Kühne's violet, which we will consider later, the iodine solution of Gram may be replaced by the following:

Corrosive sublimate, . . .	1 gram.
Water, ...	100 grams.

Alcohol, enough to dissolve.

All microbes do not take indifferently the colorations obtained by the above methods. For a given germ one method may be more suitable than another; the manner in which the microbes behave toward the different methods of staining is, indeed, utilized in distinguishing them. But it is often desirable first of all to determine whether or not a given substance contains any germs. We attain this end by the following method:

Kühne's method.—Kühne has succeeded in staining nearly all microbes by employing two processes in succession, one with methylene blue, and the other with crystal violet. Some microbes are colored only by the blue (typhoid bacillus), others only by the violet (bacillus tuberculosis, bacillus of leprosy); others, again, and these are the most numerous, take the two colors indifferently.

This method is therefore useful for determining the presence of microbes in any tissue. In its execution an equal number of sections are treated by both methods.

1. *Method by Kühne's blue.* The *blue of Kühne* is obtained by adding, drop by drop, to a one per cent solution of carbonate of ammonia, a saturated aqueous solution of methylene blue until the mixture acquires a deep blue color. The sections, after dehydration in alcohol, remain ten minutes in this blue liquid, are then rapidly decolorized (two or three seconds) in a 1 to 1000 aqueous solution of hydrochloric acid, then passed into distilled water to remove all trace of the acid. They are afterward mounted without passing through alcohol, which would remove too much of the stain. For this end, after removal from the water, they are spread out on a slide and there allowed to dry, or better, the desiccation is hastened either by means of a current of air or very slight heating. They are then cleared in xylol and mounted in balsam.

2. *Method by crystal violet.—Kühne's violet* is obtained in the same way as the blue solution by adding several drops of a saturated aqueous solution of crystal violet to a one per cent aqueous solution of carbonate of ammonia until a deep violet color is produced. Sections, after dehydration in alcohol, are left in this solution ten minutes (two hours for the bacillus tuberculosis), washed in distilled water, and immersed for two to three minutes in Gram's iodine solution; from this they are transferred into a saturated alcoholic solution of fluoresceine until decoloration is nearly complete. The remaining coloring matter is extracted

in absolute alcohol, the sections are cleared in oil of cloves, passed into xylol, and finally mounted in balsam.

To obtain double staining by the violet method it suffices to color the sections with picro-carmine before placing them in the violet solution.

Kühne has modified this process so as to avoid the employment of alcohol and oil of cloves. The sections, after dehydration in alcohol, are immersed for ten minutes in the concentrated aqueous solution of crystal violet, to which has been added hydrochloric acid (one drop to fifty grams of the solution). They are washed in distilled water, treated with Gram's iodine solution, replaced in water, then passed rapidly (a few seconds) through absolute alcohol and into aniline oil in which they are decolorized. The aniline is removed by xylol, and the sections mounted in balsam. The employment of alcohol can be entirely avoided by transferring the sections after the action of the iodine on to the slide and there treating them, after dehydration, with aniline, xylol, and balsam, in succession.

Method of Berlioz.—This gives a rapid double coloration. The sections are left for a quarter of an hour in a mixture of equal parts of the two following solutions:

1.	Distilled water, ...	84 cub. cent.
	Aniline water, ...	6 "
	Methyl violet 6 B, ...	2·5 grams.
	Absolute alcohol, ...	10 cub. cent.
2.	Distilled water, ...	95 "
	90 per cent alcohol, ...	5 "
	Coccinine, ...	2·5 grams.

They are then passed through the Gram iodine solution, or a five per cent solution of sodium carbonate, through water, alcohol, etc. This method is well adapted for the staining of charbon material.

Method of staining spores in bacilli.—To stain spores dried on the cover glass they are immersed for a few minutes in hot Ehrlich's solution and washed in alcohol; the cover glass is then held for a second in methylene blue, washed again in alcohol, and mounted in balsam. The bacilli are blue, the spores are red when they have attained their development, uncolored whilst they are in way of formation.

II. *Culture of germs.*

The majority of pathogenic germs can be cultivated in appropriate artificial media placed under suitable conditions of temperature. Our aim not being to publish a complete treatise on the technique of bacteriology, we will refer the reader for details upon this point to special works on the subject, and limit ourselves to the general principles of the question.

We will review, in succession, the methods of sterilization, the composition and preparation of the various culture media, the sowing of these media and establishing them at a suitable temperature, the physical and chemical characters which the cultures may present and the preservation of these cultures.

Sterilization.

The objects and instruments which are employed in the culture of microbes must be previously freed from all living germs. Sterilization is obtained by different means: heat, filtration, antiseptics.

Heat.—As we already know, bacteria and their spores are killed by heat. For this it is necessary that the objects which contain the germs be raised to a temperature varying according to the case. There are several processes.

a) *Flaming.*—Objects which can support the action of a rather high temperature are brought into the flame of a spirit lamp and subjected, in all their parts, to a temperature of about 200°. Flaming is practiced chiefly for glassware, occasionally for metallic instruments: spatulas, platinum wire.

b) *Dry air ovens at* 150° *C.*—The objects to be sterilized are kept for about two hours at a temperature of 150°. To this end they are placed in appropriate ovens, furnished with a temperature regulator and arranged in such a manner that the temperature, through the circulation of hot air, is nearly the same in all parts of the apparatus. These ovens are protected from too free radiation by walls composed of a substance which is a bad conductor of heat, such as asbestos. Sterilization by this process is practiced chiefly on glassware, wadding and metallic objects.

c) *Koch's steam sterilizer.*—This is a generator of steam, of cylindrical form, containing water which is kept in ebullition by the combustion of gas; it is surmounted by a dome of plate tin enveloped in felt to prevent cooling. The temperature of the cylinder, charged with steam, is maintained at 100°, a temperature sufficient for the destruction of germs and spores in the moist state. Objects to be sterilized must be left in the upper part of the apparatus from two to three hours. The steam sterilizer is employed chiefly for the sterilization of culture media, and for objects which can not sustain a temperature of 150°.

d) *Papin's steamer—The Chamberland autoclave.*—Sterilization of objects in fluid media is much more rapid and certain when they are subjected during ten to fifteen minutes to a temperature of 120°. To realize these conditions the objects are inclosed in a Papin steamer, which admits of obtaining, under pressure, the temperature indicated.

Chamberland has recently devised an autoclave or steamer specially adapted for sterilization. It is now found in all laboratories. It is merely a Papin steamer modified for this particular use.

e) *Discontinuous heating.*—Some media, such as blood serum and milk, can not endure a temperature of 100° without undergoing considerable modifications. For the sterilization of these substances we have recourse to the method of Tyndal, which consists in killing the germs in full vegetation, an operation which only requires a temperature of 58°, maintained during two hours. For blood serum the temperature should not exceed 58°, for this medium becomes coagulated at a higher temperature. The first heating not having killed the spores, the substance is then brought to the ordinary room temperature, or better, to 37°; the spores then quickly vegetate and the bacteria to which they give birth will be killed by a second heating on the following day. This operation, repeated four or five times, ends in the certain sterilization of the liquid.

The apparatus employed in this method of sterilization consists of water baths maintained at a constant temperature by a regulator.

Filtration.—Germs, like all solid particles, can not pass through porous substances in which the pores are very fine. Liquids filtered through such a substance are therefore sterilized. Hence, it is possible to obtain sterilization by filtering through plaster of Paris, amianth, or porcelain.

Chamberland has constructed a filter for sterilization based upon this property. It consists of a hollow tube of porcelain, closed at one end; this tube is immersed in the liquid to be sterilized and a vacuum produced in its interior in any of the ordinary ways; the sterile liquid passes into the interior of the tube. The latter, of course, should have been previously sterilized and be free from fissures.

Antiseptic solutions.—Sterilization of instruments, anatomical specimens, the hands, etc., should be obtained by means of acidulated solutions of corrosive sublimate at 2 to 1,000, carbolic acid at 5 per cent, or creolin at 2 per cent. To obtain sterilization, the objects, after thorough cleansing, must be washed with these solutions.

Culture media.

The culture of germs in artificial media necessitates the presence in these media of all the principles essential to their life, as well as the absence of all noxious products. Each germ having special nutritive requirements of its own, the ideal would be to possess media especially appropriate for the diverse pathogenic species. A perfect culture apparatus ought to admit of the continual addition of nutriment, with, at the same time, the elimination of the residual products; but this ideal is far from being realized: the nutritive media are, for the most part, prepared empirically, starting from complex animal or vegetable products, and not only are we unacquainted with the qualitative and quantitative composition of these media, but we are also ignorant of the special nutritive value of the essential principles they contain. Moreover, our culture apparatus are very imperfect; during the progress of the culture the nutritive material becomes exhausted and charged with the products of nutrition, products which are prejudicial to the normal life of the germs. From these circumstances it results that artificial cultures do not always give individuals conforming with those which are sowed, and that often, in the case of pathogenic species, they furnish degenerated beings incapable of acting upon man or animals.

Generally speaking, all culture media ought to contain, besides the necessary nutritive substances, at least sixty per cent of water, they should be neutral or slightly alkaline, and absolutely sterile. A considerable number of media are in use; they embrace organic liquids: milk, urine, blood serum; various decoctions: bouillons of meat, hay, fruits, beer wort, etc. These materials are used alone or after the addition of supplementary nutritive substances: peptone, gelatin, glycerin, glucose, phosphates, etc.

Without entering into the details of the preparation of these diverse media we will describe those most commonly in use.

A. Fluid media.

1. *Bouillons.*

Bouillons are the media most commonly used.* They are prepared from the flesh of the different domesticated animals, but especially from that of the calf, the ox, and the chicken.

One kilogram of lean meat, free from bone, is finely minced and allowed to macerate for twenty-four hours in two liters of water, in a cool place. The reddish fluid which bathes the meat is expressed, brought up to its original volume, and to it is added one-half per cent of table salt and a trace of potassium phosphate, occasionally, also, from one to three per cent of peptone, glycerin, and glucose. The liquid is then cooked for an hour at 100° in the Koch sterilizer, by which means a certain amount of its albumen is coagulated; it is strained through linen and then neutralized with a one per cent solution of caustic potash, or a five per cent solution of sodium carbonate. The filtered and neutralized bouillon is subjected, during ten minutes, to 115° in the autoclave, then again filtered. This bouillon, introduced into a conical flask stopped by a plug of wadding, is finally sterilized by a last heating of a quarter of an hour at 115°. It is then ready to be introduced into the culture vessels. The culture vessels most commonly employed are the Pasteur bulbs. These are small vials of thin glass, flat on the bottom, and with

* [Except for special purposes, or where a large quantity of the culture is required, solid media are now more in use than bouillons.—D.]

a very short neck, which is surmounted by a ground glass hood; the latter is prolonged by a narrow tube which is closed by a plug of wadding. These vessels may be replaced by conical vials, etc., but the latter have the great inconvenience of allowing a too rapid evaporation of the liquid.

In filling the Pasteur vessels, which must previously have been sterilized, we employ with advantage a Chamberland pipette. This is a flat bottomed bulb with a long narrow tube bent to 45° and filled at one part of its length by a plug of wadding; from the side of this bulb projects horizontally a long slender tube sealed with the flame at its free extremity. The pipette having been sterilized, the lateral tube is flamed, the point broken off in the sterile bouillon and the latter slowly aspirated into the pipette. This done, a quantity (half to one centimeter in depth) of the bouillon is allowed to flow into the Pasteur culture vessels, the latter being held in the horizontal position, in order to avoid the entrance of air germs; this operation should be conducted in a room free from currents of air. The bouillon having been introduced into the culture vessels, it is advisable to test if these are really sterile by placing them during two or three days in an incubator; those in which the bouillon becomes turbid contain bacteria and consequently must be rejected.*

* [Usually the culture media, both fluid and solid, are filled into ordinary test-tubes, closed with cotton and sterilized in the steam sterilizer. Evaporation is to a large extent prevented by accurately fitting a piece of tin-foil over the mouth of the tube.—D.]

2. *Milk.*

Milk may be collected in a state of purity in sterilized tubes by introducing a sterile canula into the teat after the latter has been well disinfected. This is the best means of obtaining milk free from germs, but it is one which is not generally available, and, most frequently, we are compelled to sterilize this liquid.

This is done in the autoclave at 115°. The milk, without the cream, is then introduced into the Pasteur bulbs and subjected to the test of the incubator.

3. *Urine.*

It was in urine that Pasteur first cultivated the bacteridium of charbon. The urine, after collection, is rendered alkaline, filtered, and sterilized exactly like the bouillon. It may be employed alone or after the addition of supplementary nutritive substances.

B. Solid media.

This consists of fluid media gelatinized, coagulated blood serum, potatoes, etc.

Gelatinized bouillons.

Gelatinizing substances are added to bouillon, simple or complex, so as to render it solid and transparent on cooling. The substances employed are gelatin, *gélose*, iceland moss, etc.

Nutritive gelatin.—To the bouillon whilst in preparation and before boiling is added ten per cent of gelatin. The fluid is passed through a cloth, neutralized, and the process generally conducted as described for bouillon. The product should be of a citron color and perfectly transparent after filtration. It is then introduced into test tubes plugged with wadding, and sterilized. These tubes, filled to one-third of their height, are then placed in a wire cage and kept in the autoclave at 105°, during ten minutes. The gelatin solidifies on cooling and the tubes are then ready for use.

Gélose or agar-agar.—This is a gelatinous substance coming from certain algæ of the Indian Archipelago.

It is added to the bouillon whilst in preparation, in the proportion of one to two per cent, and the further process conducted as has already been described for gelatin. The addition of the agar often renders the liquid turbid, and this turbidity persists in spite of filtration; it is with the view of obviating this inconvenience that it has been recommended to intimately mix with the mass, after cooling to 50°, the white of an egg beaten up, and then bring the whole to ebullition again. In coagulating, the albumen carries with it all the substances in suspension and the product becomes clearer. However it always remains slightly opalescent.

The filtration of this liquid, like that of gelatinized liquids in general, ought to be performed while hot, and it requires some time. A hot filtering apparatus is employed, consisting of a glass funnel contained in a larger one of copper and separated from this last by a space full of water kept at the temperature at which the gelatinous mixture becomes fluid. The filtration thus obtained is slow, for the gelatin dries and hardens upon the filter where it contacts the walls of the glass funnel.

For filtrations of this kind we have employed with advantage a special support bearing a simple glass funnel provided with a lid and adjusted above a conical vessel. The gelatinous mass to be filtered having been introduced into the funnel upon an ordinary filter, the whole is placed in the Koch steam sterilizer. We thus obtain a rapid filtration without desiccation or loss of the fluid, since the filtration takes place in an atmosphere of steam. The agar and gelatin having been poured into tubes and sterilized, the latter are placed in the cold, some in vertical position, others inclined so as to distribute the mass in a very large oblique layer; either the surface of this layer or the depth, in the vertical tubes, serves as a field for culture.

Agar has several advantages over gelatin: it is not fluidified by the growth of germs, it remains solid at 40°, and, therefore, admits of cultures in the incubator at 39°; it may be subjected to cooking for a long time without losing its gelatinizing power. On the other hand, it has the disadvantage of being always slightly cloudy, is not well adapted for plate cultures, and gives cultures which are not well defined.

Agar-gelatin.—With the aim of making plate cultures at the temperature of 39°, Jensen has recommended, for the solidification of the bouillon, a mixture of agar and gelatin. He adds to the bouillon 5 per cent of gelatin and 0.75 per cent of agar. The preparation is very clear, it is liquefiable at a temperature which does not kill the germs, and it remains solid at 39°.

Gelatinized serum.

A solid medium of frequent use is sterilized and cooked blood serum. Many methods are employed in collecting and sterilizing the serum; these we refrain from describing here and limit ourselves to a description of that which we are accustomed to use. Blood drawn from the jugular of the horse or the ox is collected in deep cylindrical glass vessels and left in the cold for twenty-four hours. The clot contracts and the serum comes to the surface; the latter is collected by means of a pipette and introduced into conical vials to be sterilized.

Sterilization is obtained by conveying the serum on several successive days into a water bath a 58°, and, each day, leaving it there for two hours. The bath is provided with a thermo-regulator by which a temperature of 58° is quickly obtained, a condition indispensable to prevent the germs from multiplying too much and, in consequence, altering the coagulability of the serum. After eight days heating the liquid may be considered sterile; it is then distributed in test tubes by means of a Chamberland pipette. Other nutritive substances, peptones, glycerin, etc., may be added to the serum before sterilization.

The cooking of the serum takes place in double walled ovens so arranged that the tubes may be inclined so as to spread out the liquid. These ovens are regulated to 70°; the tubes are removed as the serum in them becomes coagulated. Before being used these tubes are tested during three days in the incubator.

Potatoes.

Some germs admit of cultivation upon the potato. For the preparation of this medium sound tubers with smooth skin are selected, thoroughly washed in water with a brush, and allowed to soak some time in 1 to 1,000 sublimate solution. They are then cooked by steam at 100°, or in the Chamberland autoclave, cut in two with a previously flamed table-knife, and deposited in a moist chamber.

A moist chamber is usually represented by two crystallizers of different sizes, and well disinfected; a piece of filter paper saturated with sublimate solution is placed in the smaller, which is then covered by the larger. We thus obtain a moist space protected from the atmospheric germs.

Sometimes potatoes, cut in sections, are cooked in tubes ready for culture; these have a constriction near the base to retain the potato about the middle of the tube.

Methods of culture.

In order to cultivate germs we must first possess pure sowings. Occasionally the germ may be obtained pure from the organism in which it has developed, as, for example, in bacteridian charbon, chicken cholera, rouget of the pig, etc.

In most cases the micro-organisms do not occur in a state of complete purity in the parts of the organism in which they pullulate; it then becomes necessary, first of all, to separate them from other germs. We have, therefore, to describe, successively, the methods of isolation of germs, the inoculation of the various media, and the establishment of a suitable temperature.

Methods of isolation.

The isolation of mixed germs may be obtained in several different ways. One method of isolation is based upon the different properties of the species to be isolated. Thus, the germs of the septicæmias of the rabbit and of the mouse have been withdrawn from the mass of microbes which pullulate in blood in way of putrefaction, by taking advantage of the fact that these germs are pathogenic for such animals.

Pasteur has isolated the bacillus anthracis from the septic vibrio in charbonous blood in a state of putrefaction by taking into account the fact that the former is aërobic and the latter anaërobic: cultures in the air give the bacillus anthracis only; those protected from oxygen give the septic vibrio.

The influence of various chemical agents and that of heat at different degrees are also effectual means of isolation; they kill some germs or considerably interfere with their multiplication, and thus favor the predominance of other less sensitive germs.

Klebs has succeeded in isolating germs by basing himself upon their unequal distribution in the liquids in which they pullulate. Some, immobile, are found at the bottom or on the walls of the vessels; others, motile, are uniformly distributed; some again are very greedy of oxygen and are found in the superficial layer. By withdrawing germs from these various parts more of one species than of another will be removed, and by repeating the operation several times in succession one species will be obtained in a pure state.

Cohn has separated certain germs by taking advantage of the resistance of their spores to a few moments ebullition. The bacillus subtilis, or hay bacillus, is obtained by boiling neutral infusion of hay and then transferring to the incubator. The spores of the bacillus subtilis alone develop because these alone have resisted.

Germs can be isolated by more direct means.

Method of isolation by Salmonsen's capillary tube.—This author has studied, in this way, the various germs which grow in putrefactive blood. He aspirated defibrinated blood into long capillary tubes which he closed at both ends and fixed upon a horizontal card. The few germs which are included in the tube grow separately in the different parts of its length; their evolution can thus be studied and they can be collected separately for cultivation.

Method of isolation by dilution.—A material, rich in germs of different kinds, is diluted with sterilized water so that the germs are considerably rarefied. This liquid is then inoculated by drops into a series of culture vessels; those of these vessels which have received only one germ will give a pure culture on incubation.

MM. Roux and Yersin have succeeded in isolating the bacillus of human diphtheria by making with a platinum wire charged with a trace of false membrane, longitudinal streaks in a series of tubes of blood serum. The germs become progressively rarefied on the wire and ultimately the serum in the last inoculated tubes receives no more than one germ; the culture is then pure.

Method of isolation by plate cultures.—Koch has recommended a method of plate culture now practiced in all the laboratories. A trace of the substance containing the bacterial mixture is inoculated to a tube of gelatin which is then fluidified by heating at 30°. The gelatin is then agitated so as to uniformly distribute the germs throughout its substance, and poured upon a sterilized glass plate which is placed horizontally on a cold surface, and covered with a bell jar. The gelatin spreads out in a thin solid layer in which the germs find themselves isolated. The plate is then placed on a small bench in the moist chamber, in a room kept at a temperature of about 20°. Each germ gives an isolated colony, the appearance of which can be studied and from which seed can be obtained for starting new cultures. In this process the germs which grow on the surface of the layer of gelatin and which come, for the most part, from the exterior should not be taken into account.

Instead of using a plate the gelatin may be spread out in the tube itself or in any vessel presenting a very large interior surface.* This mode of procedure has the advantage over the original technique that it exempts the culture from all risk of external infection and allows of anaërobic cultures.

The substitution of agar-gelatin for gelatin consti-

* [The original "plate process" of Koch is now, for the most part, supplanted by the more convenient and safer method of Petri, in which the fluidified gelatin, after inoculation in the test tube, is poured into a shallow glass dish provided with a cover (crystallizers).—D.]

tutes another improvement, the first of these substances admitting of being raised to 39° without fluidifying. The temperature of 39° is much more favorable for the vegetation of germs than that of 20°.

Inoculation of culture media.

Inoculation of culture media should be performed in such a manner as to avoid the introduction into the culture medium of external germs, either from the air or the surface of external objects. Air germs will be avoided by considerably inclining the bulb or tube to be inoculated toward the horizontal and operating rapidly but quietly so as not to cause agitation of the air. The instruments used in transferring the seed must first of all have been disinfected. These instruments are:

1st. A somewhat rigid *platinum wire*, three to five centimeters in length, fused to the end of a glass rod of small diameter. This wire is brought to a red heat in the flame of an alcohol lamp; as soon as it is cooled the extremity is charged with a trace of the seed which is then transferred to the medium to be inoculated, shortening as much as possible the course to be traversed by the wire in order to diminish the chances of infection from the air. In the case of solid media the seed may be deposited in a line on the surface or in a vertical track into the substance of the gelatinized mass. The appearance of the culture will naturally vary with these two modes.

The platinum wire may be replaced by a needle of glass, readily obtained by drawing out a fusible glass rod over the lamp.

2nd. *Capillary tubes.*—A glass tube is drawn out in

the flame so as to reduce it to capillary dimensions; the capillary segment which has been perfectly sterilized by the heat is then closed at its two ends. When used, it is flamed, the two ends broken off, and one of them introduced into the liquid containing the seed, which ascends by capillarity: the tube is then passed into the culture medium and a drop of the fluid expelled by blowing at the other end. Such a tube can only be used once.

3d. *Pasteur's pipette.*—Pasteur has designed a special pipette for collecting and sowing, in a state of purity, liquids containing germs. It consists of a glass tube five or six centimeters in length and about one centimeter in diameter; this tube is drawn out and closed at one of its extremities the other being provided with a constriction, and filled with wadding; the whole is sterilized at 150°. In using it, the slender end is flamed, the point broken off, and the fluid containing the seed,—blood, various serosities, urine, pus, cultures, etc.—aspirated into the tube. The pipette is then introduced into the medium to be inoculated and one or two drops allowed to flow out. A liquid containing bacteria may be preserved in this tube for some time by taking the precaution to seal the capillary extremity in the flame. The Pasteur pipette can be used for sowing fluid media only; the platinum wire serves equally well for both fluid and solid media.

Placing at a suitable temperature.

Whilst most germs are able to grow at the temperature of 15° to 20° C., many of them grow better at a temperature of 30° to 40°, and a certain number are able to grow at this temperature only.

For cultivation at 20° it is sufficient to place the inoculated media in an ordinary room; this room should be heated in winter, and in summer, on the other hand, should be protected from a too free entrance of the sun's rays. If necessary, a room can be provided with double windows and padded door, in which is kept a stove receiving gas from a thermoregulator placed in the room; a constant temperature can thus be obtained.

When the cultures require a temperature of from 30° to 40°, we have recourse to special ovens. These are air ovens in which a suitable temperature is maintained.

Culture ovens are of various forms and sizes. They may have single or double walls; in the latter case the space between these walls contains a layer of water. They are, further, furnished with an arrangement which permits of the renewal of the interior air. These ovens may be regulated for different temperatures; usually they are heated to 39° C. Culture ovens have also been constructed, the temperature of which varies at different levels but remains constant for each of these.

Two principal conditions must be fulfilled in order to maintain a constant temperature; the loss of heat must be reduced to a minimum and be invariable, and the heat communicated to the apparatus must be equal to the heat lost. The first condition is obtained by surrounding the ovens with a body which is a bad conductor of heat, such as felt, sometimes by a double wall and a double door, and by placing them in a room kept at a uniform temperature. As to the second, it requires a more complicated arrangement;

the gas taken from the pipe passes first into a regulator which corrects the variations of pressure at the gasometer at different periods of the day, and then arrives at the apparatus by passing through a temperature regulator. This last is of various forms; it is usually a kind of mercurial thermometer surmounted by a chamber into which the gas enters; when the temperature rises in the oven in which it is placed, the mercury, in expanding, partially obstructs the entrance of the gas, diminishes the flow and the combustion, and depresses the temperature. This depression results in the retraction of the mercury, a larger inrush of gas, and the temperature rises again: thus the regulation takes place constantly between temperatures so close to each other that the oscillations are inappreciable.

The regulation of the ovens may also be obtained by the expansion of the water between the double walls. The water, in heating, expands, rises in a small tube, and presses against an elastic membrane of caoutchouc or thin metal, which membrane diminishes the orifice of entrance of the gas.

Aërobic cultures.—All that is needed in order to observe the development of aërobic germs is to place the inoculated media in the oven, or, as we may now call it, the *incubator*, at 37°; this incubator is so arranged that the air in its interior is continually renewed.

All the media admit of cultures being carried on at 37°. Nutritive gelatin, however, being fluidified above 25°, should not be carried above this temperature if it is desired to obtain the advantages arising from the use of solid media.

Anaërobic cultures.—Anaërobic germs, in order to grow, must be protected from the air; before attempting cultivation, therefore, it is necessary to remove the oxygen from the atmosphere and the medium in which these are expected to develop. This is accomplished in various ways. The best method consists in creating a vacuum in the culture apparatus and replacing the air with carbonic acid or hydrogen. This operation is repeated several times in order to insure the complete removal of the air. Gelatin must be fluidified before it is deoxygenated. When the operation is completed the culture vessel is sealed in the flame.

The oxygen of the air can be abstracted by sealing the culture tube within a second, containing substances which rapidly absorb oxygen, such as pyrogallic acid with the addition of a solution of caustic potash. Sometimes, also, we add to the culture media substances capable of taking up oxygen; for example, neutralized sulfate of indigo.*

* [Anaërobic cultures can also be conveniently carried on in small closed tubes completely filled with culture medium. These tubes may be filled in various ways; the following method of preparation and use is that which I have found most satisfactory: A piece of glass tubing about ½ cm. in diameter and 10 to 15 cm. in length is sealed at one end in the flame of a blow-pipe, heated throughout sufficiently for sterilization, drawn out at the other end to a thickness of from 1 to 2 mm., and sealed at a point about 6 cm. from the thicker part of the tube. The evacuation of the air in the tube is obtained by the ebullition of alcohol. A few drops are allowed to ascend (by slightly heating the tube, breaking off the point and immersing it in the fluid), shaken to the bottom and the tube then held in the flame of a Bunsen burner. Long forceps, the points of which are wrapped in asbestos, are most suitable for this purpose. The essential points in this process

Physical and chemical characters of cultures.

Physical characters.—When germs proliferate in artificial media their cultures assume characters which vary with the nature of these germs and with the media.

In cultures on gelatin plates the colonies assume different aspects. Sometimes these colonies are rep-

are that the heating be done carefully in order to avoid breakage and that the whole extent of the tube be sufficiently heated to prevent condensation of the alcohol vapor. As soon as the alcohol is completely volatilized the open end of the tube is sealed; an abrupt curve near the point insures its breaking at the desired place when the tubes are to be filled. A large number of these tubes can be made at one time and kept in stock. In making bouillon cultures the fluid contained in test-tubes, or (for this purpose) better, in small homeopathic vials, is boiled, allowed to cool sufficiently, and inoculated with the material from which the culture is to be made. The point of the anaërobic tube, after quick flaming, is broken off under the fluid by contact with the bottom of the vial. When the tubes have been properly made, only a very small bubble of air should be included with the fluid which rushes in. The point is then sealed in the flame, during which process a drop or two of the bouillon is necessarily expelled. By the use of gelatin or the agar and gelatin mixture, and proper dilution in the usual way, these tubes are quite well adapted for isolation of species. The colonies which develop in the substance of the solid medium never become large nor show as characteristic appearances as by the plate method, but they remain isolated and admit of pure cultures being obtained from them. In examining with the microscope or making sowings from these colonies, I break the tube near the middle, quickly flame one of the segments, and carefully heat it at the closed extremity; the cylinder of solid nutrient medium is slowly expelled and received in a sterilized Petri dish, from which the "fishing" can be performed in the usual way. In tubes prepared in this way strictly anaërobic species can be cultivated, while aërobic species, such as the hay bacillus, refuse to grow.—D.]

resented by small raised droplets of a oily appearance and of various colors: white, yellow, rose, red, purple. In some cases they appear as depressed points with regular or sinuous borders; sometimes, again, we see a felted mass of filaments radiating around a center. Certain germs fluidify the gelatin and thus produce a conical depression, full of fluid, in the substance of the medium.

When these colonies are examined under a low magnification their surface is seen to be sometimes smooth, sometimes granular, their contour regular or sinuous, occasionally bristling with filaments. All these are peculiarities which serve to distinguish the various pathogenic species.

In gelatin tube-cultures, if the inoculation has been made by puncture through the mass, the inoculation track may be seen to become turbid and gradually increase in size; sometimes the germs also grow on the surface of the gelatin and spread themselves out in such a way that the whole assumes the shape of a nail. When the growth fluidifies the gelatin around the inoculation track the space which it occupies takes the form of a funnel; but it also happens that liquefaction occurs progressively by zones from the surface toward the deeper parts. Although the inoculation is made in a continuous track the growth may only appear in isolated points; this occurs when only a small number of germs have been sown. From the inoculation made into the depth of the gelatin a series of tufts may radiate outward, thus giving the culture the appearance of a cylindrical brush or swab.

Cultures upon the other solid media may present

similar variations. As these media do not become fluidified like gelatin, they are better adapted for studying the evolution of the germ colonies.

Some germs give characteristic cultures in certain media. Glanders, for instance, upon potato, gives a slimy looking culture with chocolate brown margins; charbon, sown in gelatin by "stab" inoculation, grows in the form of a test tube brush, liquefying the nutritive medium.

Cultures in bouillon are no less variable. Sometimes we observe a uniform turbidity which is slowly deposited, in other cases clots of progressively increasing dimensions floating in the liquid; sometimes a flocculent coherent mass looking like a piece of saturated wadding suspended in the liquid (charbon); in still other cases, finally, we see agglomerated colonies floating on the surface of the bouillon like leaves of the water-lily upon water (farcy of cattle). These various appearances evidently depend upon the mode of association of the germs of which the culture is composed. The long filaments take the form of tufts of hair; chains of micrococci that of small pellets; isolated micrococci produce a uniform turbidity of the fluid.

Germs may change the color of the culture media; such changes are sometimes sufficient to distinguish these germs: blue pus. Anaërobic germs often give rise to a disengagement of gases more or less offensive, recalling those of putrefaction.

Chemical characters.—The chemical reactions to which the multiplication of germs gives rise naturally vary according to the germs concerned; they depend also upon the special conditions in which they are placed (aërobic and anaërobic). The medium becomes impoverished in alimentary substances, and, at the same time, becomes charged with excretory products. These we have already described; we will, however, repeat that they may be toxic for the germs and arrest their multiplication, and that they often communicate to the media in which they are diffused, certain pathogenic properties which the germs themselves possess.

Preservation of cultures.

The exhaustion or contamination of the media speedily brings the germs to the condition of latent life; from that time they cease to multiply; the combined action of the air and light more or less quickly destroys them. The virulence of a culture also progressively diminishes from the same cause.

This annihilation of virulence is more or less quickly produced according to the germs concerned, and also according to the composition of the culture medium. Thus, we have seen cultures of fowl cholera deprived of all pathogenic action for adult rabbits after one day, although, usually, this is preserved for sixty days, exposed to the air.

For the longer preservation of cultures of germs they may be inclosed in small sterilized tubes which are sealed by the flame so as to include as little air as possible. These tubes are kept in the dark or in wooden cases, as we are in the habit of employing. In this way their preservation is lengthened, but it is not indefinite; after a variable time, months or years, the germs die; in order to preserve the seed, it is necessary, from time to time, to invigorate the stock by making inoculations to susceptible animals, and from these obtain new cultures.

III. *Experimental infections.*

Pathogenic germs, inoculated to animals, may determine in the latter troubles of various kinds. These artificial microbic diseases are occasionally characteristic of certain germs. Nevertheless, a given germ does not always produce the same morbid conditions; these are influenced by the animal species used for the inoculation and may also vary with the individual; the method of inoculation also has its influence, and the symptoms observed may be dependent upon this method; the condition of the virus used as to virulence, and its origin (cultures or pathological products), have also an important bearing on the result of the inoculation; finally, the amount of the virus has an important influence: there are diseases which, inoculated in small doses, are inoffensive or produce immunity whilst large doses more or less rapidly result in death.

All these reasons indicate that we can draw no absolute conclusions as to the results of the inoculation, and that the latter is of no value except when combined with other means of diagnosis.

Experimental inoculations are occasionally made with the object of purifying an impure virus; in this case advantage is taken of the property possessed by certain germs of developing in a given animal species, whilst the other microbes with which they may be commingled can not live in this species.

Subjects of inoculation.—Many species of animals

may, in an emergency, be brought into use for experimental inoculation, but there are some which are preferred on account of their great susceptibility for most of the bacterial diseases and of the ease with which they can be obtained.

The animals which are the most used are the guinea pig, rabbit, rat, mouse, chicken, pigeon, and small birds. In exceptional cases the large domesticated animals are employed.

Inoculation substances.—Virulent substances used for inoculation should, as far as possible, be free from all microbic mixture, and the Pravaz syringe and other inoculation instruments should previously be rendered aseptic.

If the product to be inoculated is a bouillon culture, this fluid is turned, after shaking, into a previously flamed watch glass, and quickly aspirated into the syringe.

If the culture has been made upon a solid medium a quantity of the material is taken upon a sterilized platinum wire and diluted in a little sterilized water or bouillon.

When we have to do with fluid pathological products (blood, milk, pus, etc.) these products, collected in a pure state from the living being or from the cadaver, are employed in the way just described. In some cases they are previously diluted.

Sometimes the virulent substance is a solid pathological product. This product, free from all contamination, is crushed in a special mortar, diluted in bouillon, which is then strained through fine linen. To more completely avoid the external germs the virulent particle may be introduced into a small previously sterilized tube, and crushed by means of a flamed glass rod, the diameter of which is a little smaller than that of the interior of the tube; sterilized bouillon is then added, and, after the particles in suspension have been deposited, it is aspirated into the syringe.

Methods of inoculation.—The virus, thus prepared, may be used for inoculations performed in various ways, which we will now review.

1. *Endermic inoculations.*—This is the simplest method of inoculation; in its performance all that is necessary is to lay bare the deeper layer of the epidermis without exciting much hemorrhage, and there apply the active substance.

The hair is clipped from the region to be inoculated, a series of closely approximated superficial scarifications made with a bistoury, and the virulent substance spread over it. For this inoculation regions of the body should be selected which are not easily reached by licking, rubbing, etc.

2. *Subcutaneous inoculation.*—The object of this inoculation is to introduce the active product into the subcutaneous cellular tissue. A fold of the skin having been pinched up, the canula of the syringe is introduced at its base and the liquid to be inoculated slowly expelled. It is sometimes desirable to free the region from hair or feathers and render the point of inoculation aseptic, by the application of a strongly heated glass rod. The inoculation having been made, the canula is withdrawn, and, in order to render the result more certain and more rapid, the inoculated point is manipulated so as to lacerate the cellular tissue and accelerate absorption.

Inoculations may be made in any region, but, preferably, in places where the skin is thin and pliant (internal face of the thighs, abdominal wall, pectoral region in birds), and the cellular tissue abundant.

3. *Intra-peritoneal inoculations.*—In order to introduce the virulent matter into the abdomen, the animal being held well on the back, the abdominal walls are pinched up between the thumb and index, and the operation continued as above, after being well assured that the point of the syringe is indeed free within the abdomen. It is necessary here to operate with a certain amount of care in order to avoid wounding the viscera an accident especially to be feared in birds.

4. *Intra-venous inoculation.*—This method necessitates the employment of a liquid free from solid particles which, by their arrest in the smaller vessels, might occasion fatal embolisms. In performing this inoculation the vein is distended by pressure exerted on its course and the canula introduced into it, the point directed toward the heart; the operation is successful when a drop of blood issues at the shoulder of the canula; a graduated pressure is then brought to bear on the piston.

For this inoculation the most salient superficial veins are selected, in the rabbit the veins of the ear, in birds the vein of the arm. In some animals the inoculation can be made into the jugular, saphena, etc.

5. *Inoculations in the anterior chamber of the eye.*—The eye is first anæsthized by means of a few drops of a solution of cocaine, at 1 to 20; then, the globe being immobilized, the canula is insinuated horizontally through the cornea in its eccentric part. One

or two drops of the liquid are then expelled and may be seen diffusing themselves in the transparent media.

. 6. *Intra-cranial inoculations.*—It is necessary here, in the first place, to trephine the cranium, an operation which necessitates a variety of instruments. In small animals, however, the operation can be very simply performed. The animal being made fast, an incision is made in the skin over the frontal bone outside of the median line, the periosteum is crucially divided and, with the point of a strong bistoury held vertical to the surface of the bone, a small opening is made by rotating the instrument upon itself.

When the opening is considered to be of sufficient size the canula, thoroughly sterilized, is introduced under the cerebral envelopes and a few drops of the liquid injected.

Inoculations may also be made into the various serous cavities, the trachea, muscles, etc., but the technique of these operations needs no special description.

Besides these contagions by inoculation, an organism may also be artificially infected by the methods of ingestion or inhalation of virulent products. These methods give results relatively less certain.

Collection of virulent products.—Pathological products can be collected either from diseased animals or from their cadavers.

On diseased animals the process differs according as it is desired to collect a liquid (blood, pus, etc.) or a solid particle. Great precautions should always be taken in order to avoid the common germs which surround us. After the region has been shaved it must be sterilized by means of a strongly heated glass rod. The solid part which it is desired to study is then extracted with flamed instruments; liquids are aspirated into the Pasteur pipette.

On cadavers the same precautions are necessary. It must be remembered that, in contagious diseases, decomposition usually proceeds very rapidly and, in consequence, the organism soon becomes invaded by the germs of putrefaction. The cadaver having been fixed in the proper position, the autopsy must be conducted in a methodical manner and the products, solid or liquid, which it is desired to study must be collected aseptically.

It should not be forgotten that the disease in question is a contagious one, and that no part of the animal should escape the destruction which ought to follow all autopsies of this kind; it is well, also, to remark that several microbic diseases are transmissible to man; the operator, therefore, should take the precaution to protect himself against infection.

Solid products, when collected, should immediately be used for inoculations and for cultures, or should be inclosed in sterilized tubes. Liquids should be sealed in the Pasteur pipettes in which they have been collected.

PART THIRD.

MICROBIC DISEASES INDIVIDUALLY CONSIDERED.

I.

Microbic diseases consecutive to wounds.

Surgical wounds, treated strictly according to antiseptic rules, are protected from pathogenic germs; they heal without excessive swelling and without suppuration or fever. The healing process is limited to the extent of cell proliferation necessary for reparation only. It is the same with accidental wounds when they are rendered aseptic; but if, by the contact or the subsequent action of the object causing the injury, or by contact with clothing of any kind, harness, litter, water, or the atmosphere, pathogenic germs are introduced into the wound, diverse pathological changes may be observed.

We have seen, in the general part of this work, that a number of pathogenic germs, for example, those of suppuration and septicæmias are encountered almost everywhere; we meet with them also in most of the morbid conditions which complicate wounds; these diseases we shall first of all consider.

The lesions of a microbic nature which develop consecutive to wounds are local, remote, or general.

Local lesions consist of inflammatory processes the predominant character of which depends principally on the special pathogenic property of the microbes contained in the wound; these processes may be essentially exudative; *inflammatory œdemas, erysipelas;* or suppurative: *superficial suppuration, abscess, phlegmon;* or hypertrophic: *actinomycomata, botryomycomata, anatomical tubercles:* or, finally, gangrenous: *traumatic gangrene, diphtheria, hospital gangrene.* The dominant character of these inflammations depends principally, but not exclusively, on the special nature of the germs distributed upon the wounds; in reality, the reactionary powers of the tissues have also their influence here, and a given germ, the streptococcus of erysipelas for example, will occasion sometimes a simple dermatitis with interstitial and superficial exudation under the form of vesicles, phlyctenae (erysipelas), sometimes it will determine, in addition, a deep seated suppuration (phlegmonous erysipelas), and even necrosis of the inflamed tissues (gangrenous erysipelas).

Remote lesions manifest themselves in the organs in *direct* or *vascular* continuity with the tissues originally attacked. Hence, we see peritonitis occur consecutive to changes of a microbic nature located in the abdominal viscera: wounds of the intestine, of the uterus, metritis, etc. In respect to vascular continuity, this involves alterations of the corresponding lymphatics and blood vessels (*lymphangitis, adenitis, phlebitis, endocarditis, thrombic and embolic lesions*).

Finally, as general troubles, we have *pyæmia* and *surgical septicæmias.*

Suppuration.

We find in pus, besides the cells which constitute its essential part, various species of microbes. Their constant presence in suppurative inflammations led to the supposition that suppuration was only produced through the agency of microbes. This theory was further supported by the results of experimentation; thus, subcutaneous injection of irritating substances previously rendered aseptic caused an inflammation corresponding to the irritating power of these substances, but a non-purulent inflammation. Diapedesis of the white corpuscles of the blood seemed therefore to be dependent upon the presence of microbes.

Similar experiments, however, have led to contrary results in the hands of other investigators, and suppuration has been produced by means of chemical substances (croton oil, oil of turpentine, silver nitrate, mercury, cadaverin, etc.) without the intervention of bacteria. Moreover, it has been found that the sterilized cultures of the staphylococcus pyogenes as surely determine an abscess as the staphylococcus itself; the pus of this abscess, however, is not itself pyogenic. Substances with special pyogenic properties have, in addition, been extracted from cultures of the bacilli of glanders, tuberculosis, charbon and of Friedländer's pneumococcus. Thus is explained the possibility of spontaneous abscesses the pus of which is free from microbes; the latter, in such case, are present in another part of the economy and the pyogenic substances which they secrete being absorbed, we can understand that they may determine purulent inflammation of a predisposed organ or tissue.

These considerations, important as they are from a theoretical point of view, have, in practice, only a secondary interest. Under natural conditions suppuration is really always the result of microbes, acting not of themselves, as was at first supposed, but, according to recent researches, by means of the toxines to which their nutritive exchanges give rise.*

The rational application of antiseptics in the dressing of wounds has, further, to a large extent, demonstrated this truth, in making union by first intention the necessary termination of operative wounds.

The most common bacteria of suppuration are:

the *staphylococcus pyogenes aureus* (yellow pus);
the *staphylococcus pyogenes albus* (white pus);
the *staphylococcus pyogenes citreus;*
the *streptococcus pyogenes,*

and several other species, including a bacillus—the *bacillus pyogenes septicus.*

These are met with in the various suppurative processes: phlegmons, abscesses and the effusions of pyæmia, purulent inflammations of the external and internal surfaces, etc.

The yellow staphylococcus has also been found in furuncles and malignant pustule; the particular characters of these diseases seem to depend on the mode of penetration and the localization of the pyogenic germs rather than on the special nature of the latter. We have ourselves encountered the staphylococcus albus in furuncles which had developed in large num-

* [According to Buchner's investigations, the pyogenic property of sterilized cultures of many bacterial species resides in the bacteria themselves, and not in the chemical products which they secrete.—D.]

bers on the back of a colt a few weeks old, and which was successfully treated in the college hospital.

The streptococcus pyogenes seems to be also the causative agent in *erysipelas* and *puerperal fever* under its various forms; at least, the streptococcus of these two diseases can not be satisfactorily differentiated from that which we are now considering.

Fig. 2.

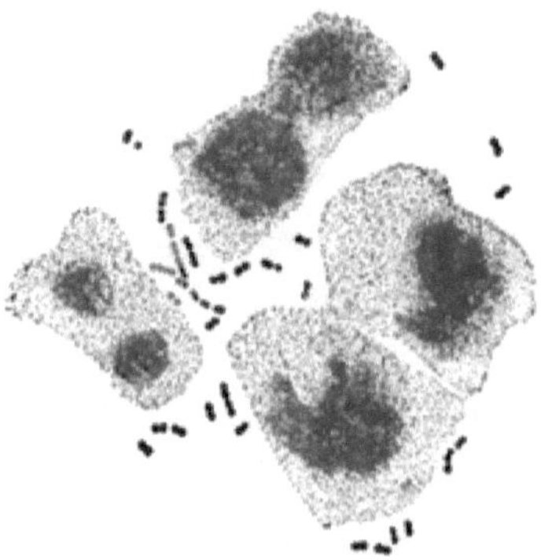

Streptococci in pus of closed abscess of horse. × 1,000.—D.

The streptocooccus pyogenes is widely distributed and its virulence is subject to great variations; when injected to animals, and especially to the rabbit, it most frequently gives rise to a local abscess, but it can also, like the staphylococcus pyogenes, occasion multiple abscesses in the muscles, kidney, lung, etc., and more or less quickly bring about a fatal termination. The staphylococcus aureus is encountered still more frequently than the streptococcus.

Cultures.—The pyogenic germs are easily cultivated in the different media; the three staphylococci are distinguished from each other by the color of their cultures, the *aureus* giving a golden yellow color, the *albus* a white color, and the *citreus* a citron yellow color.

The staphylococci pyogenes are aërobic; the streptococcus is rather anaërobic.

The chemical study of cultures of the staphylococcus aureus has led to the discovery of a non-toxic ptomaine, a non-nitrogenous phlogogenic and pyogenic substance called *phlogosine*, and a diastase also pyogenic. The yellow coloring matter is only developed in contact with the air.

Research and coloration.—In order to bring into view the microbes of pus it suffices to stain the latter on cover glasses for a few seconds with a hydro-alcoholic solution of gentian violet. The Gram reaction, which they all sustain, admits of obtaining double coloration. We, by this means, see the microbes between, and often even inside, the pus corpusles.

Etiology.—The germs of pus most frequently penetrate into the tissues through a solution of continuity, but they may also invade the organism when the skin is intact; this has been demonstrated by Garré; frictioning of the arm with a culture of staphylococcus aureus caused, on the same day, the eruption of furuncular pustules. Numerous examples of this truth are furnished by veterinary accoucheurs, in whom the arm becomes covered with pustules as a result of their intervention in difficult parturition or in removal of the placenta. In this case the lesions always develop at a pilo-sebaceous follicle, through the orifice of which the purulent germs have entered. Finally, the germs of pus may gain the tissues by traversing the normal mucosa.

The pyogenic germs which we have just referred to are not the only ones which can give rise to the formation of pus. A number of others possessing specific virulent properties have, in addition, the power of exciting suppuration; among these are the Koch bacillus, the bacillus of glanders, the actinomyces, the bacillus of cattle farcy, the bacillus of chicken cholera in the guinea pig, the bacillus of typhoid fever, etc. These bacteria, incidentally pyogenic, have been qualified as *pyocolic;* they act by themselves or by preparing the field for the pyogenic germs properly so called. It may be further stated, in this connection, that the mere contact of pyogenic microbes with a wound or living tissue is not invariably followed by suppuration. Here, the number of microbes as well as the resistance of the tissue or of the organism to which the latter belongs, all enter into the question. It is a matter of common observation that even a simple wound in some individuals, if left to itself, will always come to suppuration. The adjuvant action of traumatisms, irritant liquids, and individual predisposition has, here, an important bearing, modifying the resistance of the tissues toward pyogenic germs.

The action of pyogenic germs can be experimentally aided by the injection of their culture products. Thus, the filtered cultures of the staphylococcus and streptococcus pyogenes contain adjuvant substances and vaccinating substances. The first are the most active and alone show their effects when these filtered cultures are injected; but they are destroyed by twenty-four hours exposure to 55° in the case of the staphylococcus, and to 110° for the streptococcus, and the injection of cultures thus heated confers immunity. According to Courmont and Dor, the adjuvant action of filtered cultures of the streptococcus endures for at least three months.

Blue pus—Pus, in the human being, occasionally exhibits a peculiar blue color; this is the result of a special polymorphic microbe, generally assuming the form of a short curved bacillus, occasionally that of a micrococcus or even of a spirillum.

The germ of blue pus is very easily cultivated; it communicates to the nutritive media a green color; by the action of chloroform it is possible to isolate form its cultures *pyocyanine* which is the blue coloring matter characteristic of the germ. The green color of the cultures results from the fact that the nutritive media have originally a yellowish color.

When injected to animals it does not cause suppuration; but its cultures are, nevertheless, pathogenic for the rabbit to which it gives a special disease, the pyocyanic disease, acute or chronic according to the dose employed, and characterized by paralysis, fever, albuminuria, and diarrhœa.

The microbe of blue pus has been encountered in our animals only by M. Cadeac, in the spleen and lymph glands of a dog killed in the last stage of lymphadenoma. The author tested its identity by culture methods and inoculations but did not succeed in transmitting the disease to the dog; he thinks that the abnormal debility of the lymphadenomic subject facilitated the installation of the germ of the pyocyanic disease.

Pyæmia.

The knowleage which we now possess of receptivity and the conditions which determine it, as well as of the situation of the pyogenic germs in the meshes of the connective tissue in cases of phlegmon, justifies us in asserting that the passage of these germs into the circulatory fluids is probably a common but generally inoffensive occurrence. If the organism should suffer from any serious disturbance, or if, in consequence of a special local alteration, these germs should penetrate in large numbers into the blood, their influence will cease to remain limited to their original focus and we will see the irruption of that formidable disease, pyæmia. This disease is not absolutely dependent upon the existence of a wound; it can originate in the course of a purulent visceral inflammation or even appear spontaneously. In this last case the microbes come either from the mucous membrane or from an old latent focus.

Besides the immediate general phenomena, such as those of a peculiarly intense fever, the arrest and multiplication of the pyogenic germs in different parts of the circulatory apparatus lead to the production of metastatic abscesses and purulent collections in the natural cavities. The pyogenic germs communicate to the blood corpuscles a certain degree of viscidity; the latter become agglutinated and, as a result, embolisms occur in the small arterioles of one or more organs: kidney, liver, lungs, muscles, etc. The microbes arrested at these embolisms form the starting point of so many foci of suppuration.

In reality pyæmia is rarely simple; along with the bacteria of pus the blood generally receives other microbes from wounds exposed to the air, and there results a concomitant disease of a septicæmic order. We shall see, further, that the organisms of pus may, under certain circumstances, give, of themselves, a pure septicæmia.

Septicæmia.

In a general way this name is given to the pathological condition which follows penetration of putrid matters into the system. We now know, however, that putrefaction is only a complex fermentation tending to the degradation of organic matter, with or without the production of fetid odor, a process which necessitates the intervention of germs or microbes of various species. Hence we can not conceive of septicæmia without microbes.

But the rôle of the latter has not always the same importance; the germs, remaining intrenched outside the tissues upon a wound, for example, act solely through the soluble products (ptomaines and diastases) to which their nutritive exchanges give origin, and we then have to do with a kind of poisoning; or, the same germs, breaking the barrier which the living tissues oppose to them, make their way into the latter and multiply there; intoxication by toxines then becomes complicated with troubles caused by this proliferation.

The first of the two contingencies which we have just mentioned occurs under a multitude of conditions. And, first of all, it is unquestionable that septicæmia sometimes can be produced without the presence of microbes. Koch has shown that five drops of a putrefying fluid kills a mouse in a few hours, no microbes then being found either in the blood or viscera. The disease thus developed is a true intoxication by soluble microbic substances or ptomaines engendered outside of the organism in the putrefying fluid. It is to this species of septicæmia that we must refer the intoxications which result in animals and man from the consumption of imperfectly preserved foods in way of decomposition or already decomposed; such, for example, as the poisoning (*botulisme*) which occurs in the human species from the use of so-called Boulogne sausage prepared by unscrupulous dealers from the flesh of animals which have died or been slaughtered, diseased or sound, but nearly always, under the mask of seasoning, already abandoned to the sway of microscopic life which not only uses up its substantial parts, but elaborates diverse ptomaines of formidable toxic power.

At other times septicæmia is a sort of auto-intoxication, the septic poisons being elaborated within the economy itself but upon a limited surface which the microbes do not break through, either by reason of their special properties (the anaërobes are unable to live in the blood during life), or because the toxicity of the products secreted and absorbed is such that death supervenes in too short a time to allow of the invasion of the circulatory fluids; this surface is most frequently the seat of a morbid microbic process: for example, a septic wound, puerperal metritis, gangrenous pneumonia, etc.

But the germs of the intestine may give rise to troubles even in the absence of special alterations of the mucosa. This takes place when their products of denutrition—ptomaines, indol, skatol, gaseous products, etc.—instead of being eliminated by way of the rectum, are absorbed by the blood. The chemical poisoning which then results has received the names *stercoræmia* and intestinal septicæmia. This poisoning supervenes under a number of pathological conditions, especially when the gastric juice is insufficient in quantity or in acidity to neutralize the majority of the germs which pass through the stomach, and, in a general manner, in all somewhat lengthened diseases which lead to a prolonged retention of fæcal matters in the intestine (fever, inappetence, retention of bile, etc.). The indication, in these conditions, is to retard the intestinal fermentations by the administration of special antiseptics, and to assist in the evacuation of the toxic substances and the germs which produce them, by means of purgatives. This evacuant action of laxatives explains their utility in all febrile affections in which, for special reasons, these remedies are not contra-indicated.

The pathogeny of septicæmia, therefore, is not of one kind only. The nosological extent of the term septicæmia is, moreover, very imperfectly defined. When we take into account the fact that bacteria contribute in a general way to the reduction of complex organic molecules to formulæ more and more simple, and, on the other hand, consider the difficulty of determining, in this immense work of microbes, what should be regarded as putrefaction, and what definition should be given of putrid matters, we must recognize that, from the mode of action of the germs which occasion them, as well as by their evolution, all general diseases of a microbic order ought to be included in the group of septicæmic affections. If a certain number of these have been placed apart, this is on account of the specific characters of their germ which, giving to the disease a special expression, have established for it a well defined morbid entity: charbon, typhoid fever, tetanus, etc.

Prof. Degive had already twenty years ago conceived and clearly expressed this manner of view in several reports and discussions.(1)

He recognized in these diseases actual points of consanguinity which led him to refer them all to the same stock and to regard them as children of the same family, "The *septoid* stock or family, also called *septicæmia*, but better designated *septose*, the *septic* or *putrid* diathesis." If the discovery of the special germs of most of these diseases no more allows of belief in their identity, it removes all doubts as to their analogy.

The name *sapræmia* has been reserved for septicæmias engendered by microbes which develop a strong odor of putrefaction.

Septicæmia was formerly considered only as a complication of wounds, but it can occur independent of all traumatism, the agents which occasion it penetrating by one of the natural surfaces without the aid of any solution of continuity, as, for instance, by the digestive, respiratory, genital, urinary passages, etc. There are, therefore, *surgical septicæmias* and *medical septicæmias.*

When septicæmia develops in consequence of a traumatism we see in the latter important changes supervene which indicate the infection of the wound by germs: the secretion becomes sero-sanguinolent, often fetid, the granulations become less firm, pultaceous, often purple; the neighboring tissues swell and become the seat of an inflammatory œdema, frequently progressive; these local lesions may entail

(1) See Annales de médicine Vetérinaire, 1874, p. 502; 1875, p. 94, and 1876, p. 115.

very serious disorders, such as gangrene and detachment of the adjoining tissues; sometimes an actual putrefaction establishes itself in the living animal on the part invaded by the germs. We can not, however, assign any fixed rule as to the importance of the local troubles; they manifestly depend upon the nature of the germs which occasion them.

As to the general symptoms, they are the consequence of the absorption of the toxic substances elaborated by the micro-organisms at the wound itself. The appearance of febrile symptoms, and more especially abnormal elevation of temperature, are the first indications of the intoxication; they enable the practitioner, in the absence of any other evident cause, to assure himself of the infection of the wound, and they constitute an important indication for its subsequent treatment. Along with this, poisoning of the nerve centers by the bacterial ptomaines shows itself by the phenomena of coma, stupor, or even delirium.

The blood corpuscles, in contact with these substances, undergo a more rapid destruction, and their coloring matter, in excess in the plasma, communicates to the interstitial fluids, and notably to the visible mucous membranes, a dull yellow tint more or less intense, which, in severe cases indicates the existence of a kind of hæmaphic icterus; the production of this last condition is facilitated by the circumstance that the liver, itself altered, is unable to eliminate the coloring matter which the blood plasma carries to it in excessive amount.

The germs of septicæmias are not always found in the blood; some of them are anaërobic and unable to multiply there; in the case of others they may be seen to penetrate into the circulatory fluids when the disease is of rather long duration. It is evident that this penetration can be followed by new troubles localized in the different parts of the economy in which these germs become arrested. But even in the absence of microbes in the circulation, general symptoms of septicæmias are nearly always complicated with local symptoms due to the irritant action of the absorbed ptomaines. Such is the pathogeny of parenchymatous nephritis with albuminuria, hepatitis, enteritis with diarrhœa, pneumonia, pleurisy, pericarditis, myocarditis, endocarditis, meningo-encephalitis, etc., which are seen to develop in the course of septicæmias.

Septicæmic fever presents several varieties, the clinical importance of which should not be disregarded. *Traumatic fever* is the mildest form, the first degree of *acute septicæmia;* the latter develops in several days and determines a greater and more lasting elevation of temperature; *super-acute, fulminating*, or *gangrenous septicæmia* kills the subject in a very short time; finally, *chronic septicæmia*, or *hectic fever* occasions a slow pining away of the patient, and is of much longer duration.

Septicæmia may succeed to a large number of local lesions: sanious wounds, abscesses, erysipelas, furuncle, gaseous gangrene, etc. The microbes which occasion it are very variable and the symptoms are therefore not always alike.

The pyogenic germs can also give rise to a pure septicæmia; the streptococcus pyogenes is the cause of puerperal fever; the staphylococcus pyogenes aureus has been met with in the blood in several cases of septicæmia. In such cases the virulence of these germs is generally very great and death supervenes too quickly to allow of the formation of pus. We thus see puerperal fever under three different types in which the streptococcus pyogenes is always found; it sometimes assumes the form of a true septicæmia quickly leading to death; at other times the patient succumbs with an abscess of the large ligaments and generalization of the streptococcus, without occasioning new abscesses; finally, the disease sometimes evolves comparatively slowly, assuming the characters of a pyæmia with multiple streptococcus abscesses.

These observations show that in septicæmia and pyæmia the process is essentially the same; the result, in case of pyæmia, proceeds from the special pyogenic property of the germ and from its particular degree of virulence.

The lesions found at autopsies of septicæmic subjects are far from being constant. However, it is observed that the bodies rapidly putrefy. The parenchyma of the liver, kidney, and spleen are often inflamed and softened. We may also find inflammation of the various serous membranes: pleura, peritoneum, pericardium, endocardium, etc.; multiple hemorrhages and the more or less icteric color of all the tissues indicate, in certain cases, profound alteration of the blood.

Pasteur's septicæmia.

We will not delay to describe the history of the septicæmias experimentally obtained; they present, indeed, great interest from a general bacteriological point of view, but, here, we will only study those germs which the veterinarian is liable to meet with in practice and which it is absolutely necessary that he should be acquainted with.

The *bacillus septicus* or *septic vibrio* takes first rank among those which give rise to the most characteristic phenomena of septicæmia. It is the cause of the complications of wounds described under the names of *gaseous gangrene*, *fulminating gangrene*, *traumatic gangrene*, and *malignant œdema.* The disease which it occasions in man and animals has been designated *gangrenous septicæmia* by MM. Chauveau and Arloing.

Characters of the septic vibrio.—It is a rod measuring 4μ in length by 1μ in breadth, hence shorter than that of bacteridian charbon; it is often jointed like the latter but its different segments have not all the same length, whilst the articulated segments of charbon are of uniform dimensions. Further, whilst the segmented bacteridia are cut at right angles and slightly swollen at their ends, these characters are not observed in the septic bacillus. According to Chauveau and Arloing, when examined in the œdema of a septicæmic focus, it shows itself: 1st. with the characters of a bacillis (6μ to 50μ by 1μ to 1.5μ) provided with a spore at one of its extremities, which is occa-

Fig. 3.

1, 2, 3. Bacilli of Pasteur's septicæmia in the peritoneum and in the blood.

4, 5. Segmented bacilli.

6. Bacillus with terminal spore.—M. and L.

sionally swollen; 2d. or those of a bacillus with homogeneous protoplasm, a little longer than the preceding (12μ to 30μ). In the serous membranes, and in the blood after death, it grows to a considerable length and more or less rapidly becomes segmented into articles of varying lengths, never sporulated after the manner of the bacillus of charbon.

The septic vibrio is absolutely anaërobic; it exhibits very active flexuous movements which are quickly arrested by contact with oxygen.

Action of physical and chemical agents.—Heat is the surest and most active agent of destruction of the septic vibrio and the only one to be recommended in practice; virulent serosity is rendered inoffensive by heating for fifteen minutes at 100°; dried serosity requires a little less than ten minutes at 120°. It slowly loses its virulence through the influence of putrefaction (in two months); the virus, dried at temperatures varying from 15° to 38°, is indefinitely preserved. Antiseptics have a feeble toxic power for the septic vibrio; according to MM. Chauveau and Arloing sulphurous acid has shown itself the most powerful; sublimate, at 1 to 500, does not kill the bacillus of gangrenous septicæmia. Carbolic acid, at three per cent, is only efficacious when supplemented by heat.

Cultures.—The bacillus septicus multiplies in all the artificial media under the express condition that these media and the atmosphere in which they are inclosed are deprived of oxygen. The development is accompanied with the disengagement of carbonic acid and hydrogen; bouillons become turbid and then clear by the deposition of the bacilli; gelatin is fluidified. Agar is well adapted to its culture; on this medium it forms a whitish track with festooned borders and extends through all the nutritive mass through the breaking up of the latter by the liberated gases. In bouillon the bacilli are undulated or straight; at first homogeneous, they later become granular and break up. Some of the elements become sporulated; the spore most frequently appears at a swollen extremity of the bacillus and gives to the latter the appearance of a bell clapper. When cultivated upon solid media the bacilli are shorter, and fructification is more delayed.

Inoculation of culture media should be made either from the peritoneal serosity or from the blood of septicæmic subjects. But as the blood contains very few germs immediately after death, it is necessary, for the inoculation to be successful, to allow them to multiply there. For this end a little blood is inclosed in a pipette and kept in the incubator for twenty-four hours; multiplication of the bacilli takes place and soon shows itself by the appearance of bubbles of gas. The blood may then be employed for the inoculation of artificial media. Muscle juice can be used for the same purpose.

Research and coloration.—Examination should be made of the serosity of the œdema or of the septicæmic focus, of the blood, peritoneal serosity, and muscle juice. By studying fresh unstained preparations we can appreciate the movements of the bacilli which wind among the elements of the blood or serosity; it will be noticed that the rods in the center of the preparation preserve their motility much longer than those in the vicinity of the margin of the cover glass where they are killed by contact with the air.

Experimental inoculations.—The species endowed with receptivity for the bacillus of gangrenous septicæmia are: *guinea pig, rabbit, sheep, goat, horse;* then in the order of decreasing affinity, the *ass, chicken, pigeon*, and, finally, the *dog* and *cat*. *Cattle are absolutely refractory;* this is a remarkable exception which, itself, is sufficient to differentiate this pathogenic agent from that of symptomatic charbon, with which is has several points of resemblance.

Inoculation is most successful when made in the subcutaneous cellular tissue (one-fifth of a drop to five drops of the virulent serosity are sufficient); it invariably fails when made with the lancet or by superficial scarifications. When injected into the circulation the animal can tolerate doses much larger than those which prove fatal when injected into the subcutaneous cellular tissue. Subjects inoculated in this way suffer from a fever of more or less intensity and obtain immunity; but, when the quantity introduced into the blood has surpassed one to three drops of the virulent serosity in the rabbit, one to five cubic centimeters in the sheep, ten to thirty-five cubic centimeters in the ass, it leads to the death of the inoculated subject. In animals which have been rendered refractory by intra-vascular injection later inoculation in the connective tissue produces a slight swelling or, at most, the formation of a curable abscess.

The serosity taken from the muscular tissue, the connective tissue, and the parenchymatous organs, is more virulent than that from the serous cavities.

Inoculation in the connective tissue is followed by

a violent inflammatory reaction; the region becomes hot, painful and tumefied; the engorgement rapidly extends to neighboring regions and becomes crepitant, emphysematous, in consequence of the internal formation of gas (carbonic acid, hydrogen, carburetted and sulfuretted hydrogen, etc.). Soon the central part becomes insensible and presents all the symptoms of mortification. Gaseous infiltration may be wanting when death supervenes too quickly. The latter arrives at the end of twelve to fifteen hours in the guinea pig. The autopsy shows great detachment of the tissues, their infiltration with a sanious fluid and fetid gases, gangrenous patches of greater or less extent, and often great extension in all directions of the original inflammation.

Intra-vascular inoculation may be followed by the same changes when a solution of continuity of the circulatory apparatus allows the germs to penetrate into the connective tissue and multiply there, sheltered from the oxygen of the blood.

The disease is transmissible from the mother to the fœtus.

Etiology and pathogeny.—The septic vibrio is almost every-where present—in the soil, the dust of hay, in most putrid substances and also in the digestive canal of healthy animals. But in this last situation it is inoffensive; after death, however, the oxygen becoming deficient in the tissues which during life oppose themselves to its penetration, it invades these tissues, multiplies there, and in this way we are able to establish its presence in the blood (first of the portal vein, then throughout the economy) and on the surface of the abdominal viscera. Similarly, when it accidentally penetrates into the living tissues and finds there good conditions of vitality, it only multiplies locally and not in the blood, the oxygen of which destroys these microbes as soon as they enter this medium; on the contrary, some time after death it is found in the circulatory fluids.

Large wounds, exposed to free contact with the air, are not easily contaminated on account of the anaërobic character of this bacillus. On the other hand, the latter readily implants itself in irregular contused wounds where the affected tissues are in way of necrosis. The following experiment of MM. Chauveau and Arloing is decisive in this regard. If, after the injection of a few drops of the virus into the jugular of a ram, the circulation is arrested in one testicle by *bistournage*, this testicle becomes the starting point of a fatal gangrenous process.

When a wound becomes contaminated with the septic bacillus, if the local conditions are not adverse to its multiplication, an inflammatory process is seen to supervene the characters of which are those of the experimental œdema noted above. Absorption of the products elaborated by the bacilli, in other words, the septic intoxication, leads to general manifestations of the disease which, sooner or later, terminates fatally. In fact, when 30 to 40 cub. cent. of the serosity of the œdema, deprived of its bacteria by filtration, are injected into the guinea pig, it leads to death in a few hours with symptoms of septicæmia.

Transmission of the disease from one subject to another most frequently occurs by means of surgical instruments which have been contaminated by contact with an infected wound; this is the cause of the epidemics which have been observed in mankind; since the time that disinfectants came into general use this complication of wounds has become extremely rare.

Septicæmias of the rabbit.

The rabbit is very sensitive to the action of pathogenic microbes; it succumbs in a comparatively short time to inoculation with the majority of these germs, and for this reason constitutes an important laboratory auxiliary. The experiments of Davaine, Coze and Feltz, and of Koch upon the virulence of putrid material were made principally with this rodent. Of these researches we have only to refer to those which relate to the experimental septicæmia of Koch; the microbe peculiar to this disease shows, indeed, many characters in common with those of certain diseases of our animals, such as chicken cholera, duck cholera, pneumo-enteritis of the pig, and the epizootic of deer (*wild-seuche*). But, independent of all experimental conditions, rabbits may contract diseases of a similar nature, which prevail epizoötically in their hutches and occasion serious losses. These, in contradistinction to the preceding, are designated *spontaneous septicæmias.* We will study here, in their essential points, first, the experimental septicæmia of Koch, and then the spontaneous septicæmias. The reader will readily notice the points of similarity of these different diseases.

Experimental septicæmia of the rabbit. (*Koch.*)

Koch produced this disease by the injection of a maceration of putrefied meat; he obtained a putrid phlegmon, and death at the end of three days; the œdematous serosity at the periphery of the abscess, when inoculated in very small doses, transmits a septicæmia, fatal in twenty-four hours, to successive series of rabbits.

The œdematous fluid and the blood contain, in large numbers, ovoid microbes $0{\cdot}8\mu$ to 1μ in length, of which the extremities only take the color, so that, after staining, it affects the form of an 8.

The disease is easily transmissible by inoculation to all species of birds; according to Petri it may even prevail as an epizootic in chickens, ducks and geese. It is not transmitted to the guinea pig.

In the rabbit it gives rise to the following lesions: œdema at the point of inoculation, hemorrhagic patches on the peritoneum and in the lung, and enlargement of the spleen. Inoculated birds show a rapid emaciation with lowering of the body temperature; death, preceded by convulsions, arrives in less than twenty-four hours. The alterations consist in ecchymoses in the cellular tissue, abdominal effusion, petechiæ upon the intestine, infiltration of the lung, and the presence of a spumous mucus in the bronchi. The blood of birds transferred again to the rabbit in the smallest traces, reproduces the septicæmia in the latter.

Spontaneous septicæmias of the rabbit.

Lucet has described a disease of these animals which prevailed in the hutches and occasioned serious losses. The subjects show inappetence, emaciation, torpor and temporary muscular spasms; diarrhœa also occasionally supervenes, and death always quickly ensues. At the autopsy the blood is dark, the spleen much enlarged and darkened; the pleura and peritoneum are the seat of exudative inflammation with fibrinous deposits, and there is slight abdominal effusion.

The blood and organs contain a non-motile micrococcus, isolated or associated in pairs, $0{\cdot}7\mu$ to $0{\cdot}9\mu$ in diameter; it takes the stain quite uniformly throughout, but is not stained by the Gram method. It is at once aërobic and anaërobic, quickly loses its virulence in cultures left in contact with the air, and does not vegetate on gelatin or potato.

The disease is transmissible from rabbit to rabbit, from the rabbit to the guinea pig and inversely, by inoculation, ingestion and by simple cohabitation. Its virulence becomes attenuated in the organism of the guinea pig, but regains its original strength on its return to the rabbit. In this last a culture which, from age, has lost a part of its virulence gives at first a local abscess. The disease studied by Lucet is not transmissible to the chicken. The food and dejections are the vehicles of the germ and it is by their intermediation that the natural infection occurs.

Thoinot and Masselin have also studied a spontaneous septicæmia of the rabbit which decimated the hutches at the Alfort school. The symptoms noted are loss of appetite and vigor, acceleration of respiration, and sometimes diarrhœa. The lesions consist in a dark color of the blood, deep wine color of the muscles, roseate or yellow effusion of the peritoneum and pleura, and albuminous urine.

The disease was attributed to a micrococcus, single or associated in pairs, motile, presenting in birds the appearance of a short bacillus like the figure 8, of which the two extremities, possessing more affinity for the coloring matter, are separated by a clear space. This microbe, once colored, is decolorized by the methods of Gram and of Weigert. It is found in large numbers in the blood and in most of the organs when the disease has lasted for some time; on the contrary, it is much rarer in very acute forms.

Its culture is easy; it behaves as a facultative anaërobe and reproduces itself in the form of a motile diplococcus recalling absolutely the microbe of chicken cholera. After twenty days exposure to the air, its cultures in bouillon have lost all virulence. When growing in bouillon the latter becomes turbid, then clear again; gelatin is not fluidified and the culture in this medium takes the form of a thick, white, glistening track, dentated at its margin.

The disease is readily transmitted from rabbit to rabbit; it is also inoculable to the guinea pig and to all species of birds, in which it occasions symptoms resembling those of avian cholera. When the virus is inoculated in the pectoral muscle it produces a sequestrum in all respects comparable to that caused by the inoculation of this last disease.

Chicken cholera. (*Fowl-cholera.*)

This is an extremely fatal disease which prevails epizoötically among farm-yard fowls. It is characterized by easily recognizable symptoms: the affected animals lose their accustomed vigor, generally assume the sitting posture, or move listlessly around; they are chilly and seek the sunshine, their plumage is ruffled, giving them the appearance of a ball of feathers; the comb becomes purple, bluish, or even black. The appetite soon disappears; an abundant diarrhœa ensues with expulsion of a glairy material and occasionally a viscid liquid may be rejected by the beak. Death supervenes in a few days, sometimes preceded by convulsive movements; but the disease may be fulminating, and we then find fowls dead in their nests, the disease having produced its effects during the deposition of the egg. Cholera may also be of considerable duration and give rise to a slow emaciation of the affected animals; in these cases, again, death is the usual termination of the disease.

Fowl cholera is the cause of serious losses, hundreds of poultry sometimes dying in the same yard within a few weeks. The disease appears to have not always the same degree of malignity; this peculiarity is accounted for by the great mutability of the germ which occasions it.

The changes found at the autopsy are quite constant: the blood is dark, usually tarry in appearance; however, it is not uncommon to find consistent clots in the cavities of the heart. The liver is large, dark, or sometimes of a rather light-brown color, and dotted with hemorrhagic patches; the intestines contain a quivering jelly-like mucus more or less adherent to the mucosa; the latter is inflamed, sometimes ulcerated, its alterations being more marked as the disease has been of longer duration; in acute cases we especially notice in this situation the presence of numerous petechiæ. The heart shows a characteristic lesion: its external surface is dotted with hemorrhagic points localized especially in the coronary groove; the pericardial sac contains a quantity of fluid exudate and a gelatinous deposit adherent to the heart; this deposit is rarely absent. Miliary extravasations may also be found in the nerve centers.

Fig. 4.

1. Microbes of chicken cholera in figure 8 form; 2. The same in their real form—bacilli with clear central space. (M. and L.)

Microbe. —The blood, the dejections and the pericardial exudate contain, in large numbers, a micro-organism, very short and ovoid, resembling in shape the figure 8. It is a diplo-bacterium, the extremities of which are less refringent and have more affinity for the coloring matters than the middle part, which remains clear. It often appears as a micrococcus; this is when its long axis corresponds with the direction of the visual ray and the organism is seen on one of its ends. It is motile, its movements being very rapid in preparations of fresh blood.* It measures from $0{\cdot}6\mu$ to $0{\cdot}8\mu$ in length by $0{\cdot}3\mu$ to $0{\cdot}4\mu$ in thickness. It is a facultative aërobe, oxygen, however, being favorable to its multiplication.

Action of physical and chemical agents.—The bacteria of chicken cholera are killed in fifteen minutes at 50°, in ten minutes at 80°, still more quickly at boiling temperature. By desiccation they are killed in a few days. Corrosive sublimate, at 1 to 5,000, kills them in one minute, carbolic acid, at three per cent, in six hours. They are not affected by the gastric juice. By the oxygen of the air they are first attenuated and then killed.

Cultures.—This germ grows well in bouillon in the presence of air; during its growth this medium

* [Described in German works as non-motile.—D.]

quickly becomes clouded and at the end of several days clears again by the deposition of the micro-organisms.

Gelatin inoculated in lines becomes covered along the latter by a raised transparent pellicle; inoculated by puncture it shows small gray colonies all along the needle track; it is not fluidified.

The culture succeeds very indifferently on potato.

Research and coloration.—The microbes of chicken cholera are easily distinguished, under strong magnification, in uncolored preparations of fresh blood; they appear as very refringent, mobile diplo-cocci. They must be stained with the hydro-alcoholic solutions as they are decolorized by the Gram method; Löffler's method is most suitable, especially when it is desired to study the microbes in sections.

Experimental inoculations.—The disease prevails spontaneously in and is successfully inoculated to all species of poultry: chickens, ducks, geese, pigeons, and turkeys; the pheasant and sparrow are also susceptible. Subcutaneous inoculation and ingestion give results almost equally certain. The introduction of the virulent matter into the pectoral muscle, by means of the Pravaz syringe, causes the formation of a sequestrum which is the more pronounced as the experimental disease is of longer duration. This sequestrum, however, is not characteristic of the disease; it may be seen after the injection of other germs (those of pneumo-enteritis of the pig, the spontaneous rabbit septicæmia of Thoinot and Masselin, and a number of septic bacteria).

The rabbit is extremely sensitive to chicken cholera. The subcutaneous injection of a drop of blood coming from a diseased chicken kills it in twenty-four hours. Hence, it is a delicate reagent for verifying the nature of an epizoötic of fowl cholera, but, in order that this experiment should be of real value, it is evidently necessary to carefully avoid all causes of error which may result from the intervention of foreign germs. The blood of the rabbit which succumbs under such conditions is exceedingly rich in the specific germs. The rabbit can also be infected by way of the digestive canal.

Inoculation of a drop of blood from a diseased chicken in the subcutaneous cellular tissue of the guinea pig causes an abscess which heals by evacuation of the pus; but inoculation into the blood results in death, the same as in the rabbit and chicken, by blood asphyxia. The pus of the abscess, in the guinea pig, is rich in microbes and its inoculation to chickens or to the rabbit reproduces the disease.

It appears from a number of observations that the contact of virulent blood or of a culture with a wound can occasion in man, also, the formation of an abscess, and that the cat and the dog may consume, with impunity, chickens which have died from the disease.

Etiology and pathogeny.—The germ of chicken cholera is deposited on the soil of poultry houses and yards with the fluid dejections of diseased fowls; their entrance into the organism of healthy subjects takes place by way of the digestive canal, the fowls taking up particles contaminated with the germs along with their food. The transmission of the disease by ingestion has, moreover, been experimentally proved. Since desiccation and contact with the air quickly kills the microbe, it is in moist poultry pens, in which the excretions accumulate, and in stagnant waters of the yards, that the virus is, for the most part, preserved. The germs of the disease may thus persist for a long time on one farm even after the removal of all diseased fowls.

The microbe of chicken cholera, being aërobic, extracts the oxygen from the blood and thus determines symptoms of asphyxia, which show themselves by the dark blue color of the comb, general coolness of the body, and hemorrhagic patches on the pericardium, liver, peritoneum, etc. By filtering a bouillon culture through plaster of Paris, M. Pasteur has obtained a liquid free from germs and which, inoculated to fowls, produces, temporarily, the most prominent symptoms of the disease, giving rise to a condition similar to that which succeeds to the absorption of a narcotic dose of opium. The animal is at first excited, then its feathers become ruffled, the appetite is lost, it becomes somnolent, and, after several hours, emerges from its temporary stupor. Thus we see produced experimentally the intoxication which succeeds to the multiplication of the specific bacteria in the blood.

Attenuation—Preventive inoculation.—In cultivating the microbe of fowl cholera in the air and inoculating cultures of different ages, M. Pasteur observed that their virulence progressively diminished. Thus, by inoculating each time the same number of fowls with virus fifteen days old, one month, two months and over, he observed that the mortality induced by these different specimens of virus gradually decreased, and that, at a certain time, all the inoculated fowls survived. If at the time that a fowl is inoculated with a virus of a certain degree of virulence, this same virus is inoculated to culture media, it there reproduces itself, communicating to its progeny the special pathogenic activity which it itself possessed; the attenuation of the germ is therefore, in this case, hereditary. A series of cultures of progressively decreasing virulence can thus be prepared.

The attenuation is connected with the action of the oxygen of the air; in order to preserve the virulence of cultures it suffices to exclude them from contact with the air; in contact with the latter they generally become inoffensive at the end of two months.

The attenuated virulence shows a series of degrees, from the normal virulence up to non-virulence. When the virus is enfeebled to such an extent that it no more kills chickens, it develops at the point of inoculation in the pectoral muscle a local alteration, ending in the formation of a sequestrum which may either be eliminated or absorbed. Fowls, after recovery, are vaccinated against the later action of a more virulent or mortal virus. The immunity thus conferred is the more efficacious the more intense the vaccinal disease; its duration appears not to exceed one year; the immunity from a first vaccination is strengthened by inoculation of a second more virulent vaccine.*

The enfeebled virus can regain its virulence when

* [Kitt has obtained immunity in chickens against the virus of chicken cholera by injection of the blood serum of previously immunised chickens, as well as by injection of the albumen of the eggs coming from immune hens. (*Centralbl. f. Bact. XIV*, 25.) —D.]

it is passed in succession through small birds (canaries, etc.).

Infectious enteritis of chickens.

Under the name of *infectious enteritis*, Klein has described an epizootic disease of chickens which has much resemblance to cholera; the initial symptom is diarrhœa; the subjects are quiet but never show the somnolence so characteristic of cholera. Death occurs in twenty-four to thirty-six hours after the first manifestations of the disease.

The intestinal contents, the blood and the splenic tissue contain a special bacillus measuring 0·8μ to 1·6μ in length by 0·3μ to 0·4μ in thickness. The chickens naturally become infected by ingestion of contaminated substances. Inoculation of the blood of a diseased chicken into the subcutaneous tissue of another, results in the death of the latter; but the inoculated animal remains well during the first five days and dies on the seventh to the ninth, whilst chickens similarly inoculated with fowl cholera succumb in twenty-four to thirty-six hours. Finally, the bacillus of the disease described by Klein appears not to be pathogenic for the rabbit or pigeon, thus differing from that of cholera.

Epizootic dysentery of chickens and ducks.

M. Lucet, veterinarian at Courtenay, has studied another epizootic disease of domestic fowls, which he has named *epizootic dysentery of chickens and ducks*; a summer disease as deadly as the two preceding, this dysentery especially attacks young chickens of the same year, and gives rise to symptoms which strongly recall those of cholera: somnolence, lassitude, inappetence, diarrhœa, chilliness, ruffling of the plumage, etc. The temperature, at first high, afterward descends one to two degrees below the normal figure; the animal dies on the ninth to the thirteenth day after the beginning of the disease, occasionally much later.

The bacteria, which are present in most of the lesions and especially in the intestine, show themselves in the form of short bacilli 1·2μ to 1·8μ in length, occasionally isolated, more frequently united in pairs, motile, at once aërobic and anaërobic, rapidly becoming attenuated in cultures.

The disease is communicated, by ingestion of virulent products, from chicken to duck, and inversely. Inoculation of cultures also causes it in the same fowls; but ingestion of cultures only produces the disease when the diet is changed at the same time. The pigeon and the guinea pig are refractory; the rabbit takes the disease only by intravenous inoculation; subcutaneous injection remains without effect.

Duck cholera.

Ducks are liable to contract chicken cholera; in addition, they may be attacked by another contagious disease which is also characterized by diarrhœa, emaciation of the affected subjects, and by its usually fatal termination; it has been described under the name of duck cholera.

This disease is caused by the multiplication in the blood of a bacterium presenting the greatest morphological analogies with that of chicken cholera; it is oblong, short, and appears bi-lobed in stained preparations on account of the greater affinity of its extremities for the coloring matters. The microbes of duck cholera are a little larger than those of chicken cholera. Like the latter they do not admit of double staining.

Their culture on the different solid media is richer than that of chicken cholera. This peculiarity is especially noticeable upon potato, which is a very bad field for the growth of the organism of chicken cholera, whilst that of duck cholera grows very well on this medium.

Whilst chicken cholera is pathogenic for ducks, duck cholera is inoffensive for chickens and pigeons. The latter disease does not even kill all ducks with the same rapidity; some species resist for a longer time than others.

The rabbit succumbs to duck cholera as well as to chicken cholera, but requires a larger dose of the former than of the latter.

It might be thought that the virus of duck cholera was an attenuated form of that of chicken cholera. But inoculation of the former to the chicken ought then to vaccinate against the latter disease, which is not the case.

The penetration of the germs, in the case of spontaneous contamination, takes place by the digestive canal; this mode of infection has also been demonstrated experimentally. The experimental disease can also be communicated by hypodermic injection.*

* [Several other microbic diseases of fowls have been described: Vibrio-cholera of chickens (*Vibrio Metschnikovi*, Gameleia) in Russia,—caused by an organism presenting morphological and cultural resemblances with that of Asiatic cholera; septicæmia in geese (*Spirochæte anserina*, Sakharoff) in swampy regions of Transcaucasia—a spiral microbe resembling that of relapsing fever

Bacteridian charbon.

This is an infectious and contagious disease caused by the bacteridium.*

The disease shows itself by a profound adynamic fever, with more or less marked stupor of the affected animals. The blood is much changed, viscid, and the plasma, loaded with the coloring matter of the corpuscles, communicates to the mucous membranes a dull yellow tint; sometimes visible hemorrrhages occur: nasal and conjunctival petechiæ, bleeding from the lungs and bowels, hematuria. The intestinal lesions in the horse often give rise to more or less violent symptoms of colic, and this complication, considered too exclusively, frequently interferes with the diagnosis of the essential disease.

At the autopsy the blood is found to be deoxygenated, viscid, incoagulated, the corpuscles are altered, agglutinated, and the plasma colored red. The internal tunic of the bloodvessels and of the heart is also often stained red; petechiæ are found on the heart, lung, pleura, and peritoneum. The spleen is much enlarged; its borders, clear cut in the normal condition, have become rounded; its surface is often lumpy, its consistence soft and friable, and its meshes infiltrated with extravasated blood. The intestines are

of man is found in the blood, not cultivated; epizootic disease of grouse (Klein) in England; epizootic pneumo-pericarditis in turkeys (McFadyean) in England.—D.]

* [The term "Bacteridium" or "la bactéridie," derived from the genus *Bacteridium* of Davaine's early classification is retained by French writers as a specific name for the bacillus anthracis. "Bacteridian charbon" or "charbon," refers to the disease produced by this organism. Synonyms; anthrax, splenic apoplexy, etc.—*Ger.* Milzbrand.—D.]

sometimes the seat of intense congestive and hemorrhagic lesions, and in some cases the lymphatic glands of the different regions are in the same condition, and also enlarged to twice or three times their normal size. Similar lesions may also be found in the kidneys, meninges, etc.

The species which are liable to contract the spontaneous disease are the *sheep*, *goat*, *ox*, and *horse;* this last less readily becomes infected than the first; it can consume with impunity foods which occasion the disease in the others. Charbon is also met with in the carnivora of menageries (lion) when they are fed with the flesh of animals which have died of this disease. Exceptionally the *dog* and the *pig* become infected in the same way. Algerian sheep are refractory even when they are born in other countries.

Man is unfortunately also liable to contract this disease; the latter then receives different names according to the mode of penetration of the bacteridium and the initial lesion which it determines:

1st. *Malignant pustule*, the most frequent form, consecutive to the accidental insertion of the virus into a cutaneous wound, develops in workmen who cut up, dress, or retail charbonous meat.

2d. *Pulmonary charbon*, the rarest form, develops in consequence of the inhalation of dust charged with the bacteridia, or rather with their spores, in workmen who handle wool or skins coming from the bodies of charbonous subjects:

3d. *Intestinal charbon*, consecutive to the consumption of charbonous meat.

Characters of the bacteridium.—The microbes of charbon are straight, cylindrical rods; they are isolated or associated in twos or threes, rarely more, the limits of the articles being then marked by one or more articulations or clear zones distinctly traversing the polybacillar filament; often two contiguous elements have commenced to detach themselves and form between them a large open angle with the articulation partly disconnected, forming its apex.

The different segments of the same filament are of equal length; each segment is very slightly swollen at its extremities. Their dimensions vary from 5μ to 10μ in length by 1μ to 1·5μ in thickness. The compound filaments are never of great length in the blood of animals dead of charbon on account of their constant collision with the blood corpuscles. In artificial cultures, on the contrary, they become considerably elongated, important changes at the same time being seen to take place in their substance; their homogeneous contents become modified, condensing in the form of spores; the latter are ovoid corpuscles, highly refringent, less thick than the filament itself and consequently never produce bulgings in the latter. The formation of the spores is followed more or less quickly by their liberation, through the disintegration of the filaments. Reproduction of the bacteridium, therefore, takes place only by fission in the bodies of diseased animals during life; in cultures the transversal division, only brought into view by staining, is followed by sporulation, but only between certain limits of temperature; it begins above

Fig. 5.

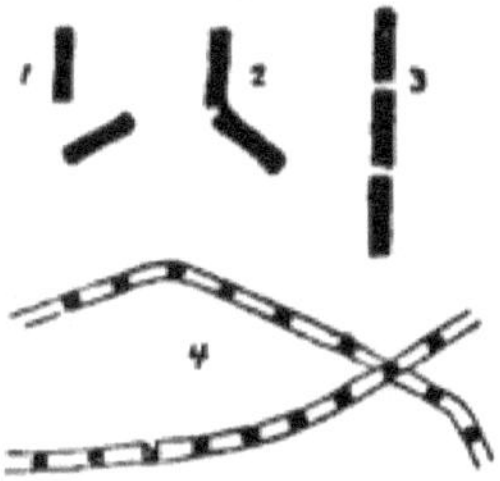

1, 2, 3. Charbon bacilli in the blood.
4. The same in a culture, sporulated. (M. and L.)

12° and ceases at 40°, and is the more active in proportion as the nutritive fluid becomes more exhausted.

The bacillus of charbon is non-motile and aërobic; the presence of oxygen is, indeed, absolutely necessary for sporulation.

Action of physical and chemical agents.—A temperature of 45° arrests the vegetation of the charbon bacilli; at 42·5° they multiply only by fission, no longer giving birth to spores, and they also become attenuated.

The rods are killed by several minutes exposure to a temperature of 50°; the spores, however, are not influenced by this temperature; in the moist condition they resist 80° and when dry support with impunity a temperature of 100°.

The bacteridia, in the condition of mycelium, that is, the rods, are destroyed by putrefaction; hence they quickly disappear from carcasses abandoned to the air. But this is not the case for those in which the viscera have been immediately removed, and more especially for bodies of charbonous animals which have been bled and dressed. Putrefaction, in this particular case, is much more tardy, and it is possible to find the charbon bacillus in the blood several days after death. This is a fact of very great practical importance and one which meat inspectors ought carefully to notice.

It would seem that, the bacilli of charbon being destroyed by the putrefaction of the cadavers, the latter would not constitute a source of danger for later contaminations. But after the death of the diseased animals all the bacilli are not destroyed; those which come into contact with the external air form spores and these are completely resistant to decomposition. In bodies from which the skin has been removed the exposure of a large number of bacilli involves a more abundant production of spores than in those which have not been skinned.

The bacilli accidentally deposited upon the soil by diseased animals, finding themselves in contact with the air and at a suitable temperature will also form spores, which are preserved for a long time on the surface of the vegetation. According to some authors, these spores themselves can pass through the different phases of their evolution in the soil and give rise to new generations.

The bacteridia retain their virulence for a lengthened period in the blood when this liquid has been carefully collected and protected from the invasion of the germs of decomposition. Dried blood also long retains its virulence; the bacilli are brought to the condition of latent life and will again multiply when under good conditions of humidity and temperature.

Antiseptics act very differently on the bacilli and on their spores. Thus one-fourth to one-half per cent solution of carbolic acid kills the rods, whilst a five per cent solution is required to kill the spores. The bacteridia are killed by carbonic acid, by compressed oxygen, and by absolute alcohol, whilst the spores resist these agents.

Cultures.—The bacilli of charbon readily multiply in artificial media at temperatures between 12° and 43°, and in contact with the air. The most favorable temperature is about 38°.

In *bouillon*, after the first day, white mucoid flakes are seen suspended in the fluid; these do not become dissociated by shaking. They slowly increase in size, still remaining united; they are formed of filaments of great length intwined with each other like the threads of cotton wadding. After a certain time these filaments produce spores and break up; the liberated spores fall to the bottom of the vessel where they look like fine sand.

Inoculated to *gelatin* by puncture, the charbon bacillus forms a culture no less characteristic; along the deep track a white line appears from which project horizontally ramifying branches, so that the whole simulates the branch of a tree with its divisions and subdivisions; on the surface a white layer is formed which slowly fluidifies the gelatin and gradually increases in thickness.

The colonies which develop on gelatin plates are formed of tufts of interlacing filaments showing arborescent prolongations at their periphery.

On *agar* the bacteridium grows as upon gelatin.*

On *potato* it produces a dry crust of a white color.

Virulent cultures contain various toxic substances: a toxalbuminoid, precipitated by alcohol (Nankin, Brieger, Fraenkel), and an alkaloid (Martin).

Research and coloration.—The charbon bacillus is found in the blood of animals which have died from the disease, and in the local œdema consecutive to accidental or experimental inoculation. It is easily recognized in fresh unstained blood preparations in the form of very clear articulated or non-articulated rods, lying motionless between the corpuscles. It

* [But the growth in agar tubes is in no way characteristic.—D.]

readily takes the different aniline stains; the blood may be stained with aqueous solutions either directly or after being dried on the cover glass; in this last case the bacteridia are generally shorter and more slender. The double stains of Gram and Weigert give excellent results.*

Fig. 6.

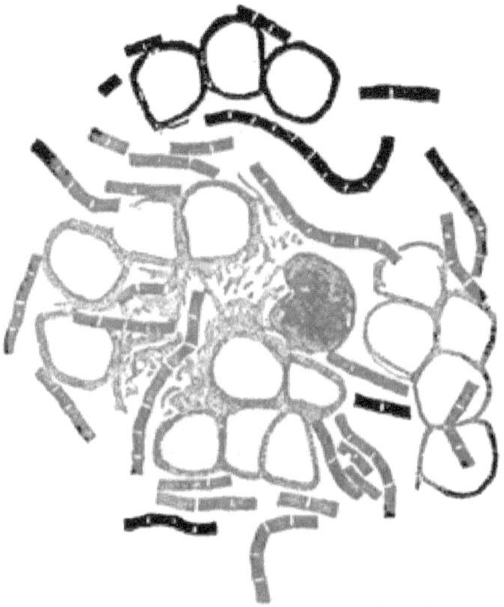

Anthrax bacilli in the blood.
(Cover-glass preparation).—Kitt.

Examination of fresh blood is especially to be recommended here, as it enables us to establish the immobility of the bacilli and the alterations of the blood corpuscles. In examining the blood of a carcass for bacteridia the specimen must be taken from a deep vein some distance from the peritoneum, because, in this disease, the invasion of microbes coming from the abdominal viscera is more rapid as the charbon bacillus, being aërobic, has deprived the blood of its oxygen.

Experimental inoculations. — Inoculations may be made by subcutaneous injection, by intra-vascular

* [The Gram stain is especially applicable to the staining of sections. On cover-glass preparations the characteristic appearance of the bacillus is best brought out by the use of aqueous or hydro-alcoholic solutions. Löffler's stain, followed or preceded by eosin, gives excellent results.—D.]

injection, or, more simply, by superficial scarifications of the integument. The disease can also be artificially communicated by ingestion, or even by inhalation. To confirm a diagnosis hypodermic inoculation and scarifications are amply sufficient.

The substances to be inoculated are represented by the blood or its serum and the diluted pulp of organs rich in vessels: spleen, liver, lymphatic glands, etc.; in laboratories cultures are also used for inoculation.

The disease is transmissible by inoculation to nearly all the domesticated animals. Mice, guinea pigs, and rabbits are the animals which are most frequently employed for experimental inoculation for diagnostic purposes. The white rat possesses an almost complete immunity, although under certain circumstances it can also be succesfully inoculated.

The *dog* takes the disease only when large doses are employed or when the virulent substance is introduced into the vascular system. *Fowls* are refractory to both the natural and experimental disease; but this immunity is easily overcome: cooling a fowl to 38°, by placing it in a current of cold water, is sufficient to induce the development of the disease after inoculation. These fowls, when again warmed, recover. The temperature of fowls is, therefore, naturally too high to admit of the pullulation of the charbon bacillus. On the contrary, that of *frogs* is too low, and we only succeed in infecting these animals with charbon when they are heated in a bath to 35°.

In the rabbit and the guinea pig the inoculation is followed, at the end of ten to fifteen hours, by a marked local œdema; the temperature is elevated from one to two degrees. Nevertheless, the appetite only fails a few hours before death. The latter supervenes in thirty-six to forty hours in the guinea pig, forty-eight to sixty in the rabbit; the animal dies in coma or after slight convulsions, and always with a marked depression of temperature.

The autopsy of animals which have succumbed to the inoculated disease is interesting: the inoculated place is the seat of a jelly-like œdematous infiltration, of a faint red color. This œdema is found in natural cases of charbon when the penetration of the bacteridia has occurred at an erosion of the digestive mucous membrane: gloss-anthrax, charbonous angina. The corresponding lymphatic glands are tumefied, ecchymosed, and surrounded by an œdematous zone. The œdematous fluid and the blood are charged with bacteridia. The spleen is much enlarged, diffluent, and dark in color.*

* [We have little reliable information as to the extent to which anthrax prevails in the United States and Canada. Local outbreaks have been reported from many different quarters, but the microscopical and laboratory investigations necessary to the identification of the disease have generally been lacking. A disease known to the natives as "charbon" prevails in some years as an epidemic among mules in certain parts of Mississippi (also reported in Louisiana and Arkansas). This disease was investigated and described as charbon by Veterinarian Creelman, of the Mississippi Agricultural Experiment Station, and the diagnosis confirmed by the laboratory investigations of Dr. Connoway, of the Missouri Experiment Station. (*Miss. Agric. Exper. Sta. Bull. No.* 11.)

Material from the same source (obtained from a cow), examined at Washington, was found to contain the anthrax bacillus. (*An. Rep. Bureau of Animal Industry*, 1889–1890, *p.* 43.)

An extensive outbreak of anthrax, in Illinois, is reported by

Etiology and pathogeny.—Diseased animals disseminate the virus in their fæcal dejections and urine, which are often mixed with blood and therefore with bacteridia; frequently, also, the blood escapes to the exterior directly by the nasal cavities. Their carcasses, all parts of which contain the virus, are more important sources of contagion; hence, when these carcasses are buried without any precautions being taken to destroy the charbonous germs, they may become the starting point of fresh cases. These bodies become more directly dangerous when the flesh is used for food.

Contagion by direct contact is comparatively rare; it occurs when a person inoculates himself with charbon at a cut or any other recent wound, in cutting up a charbonous carcase; it also occurs when the disease follows the consumption of diseased meat, and finally, by transmission, now well established, from the mother to the fœtus.

Contagion most frequently takes place by indirect contact, the germs having been previously distributed in the surrounding media. The floor of sheds, the soil and vegetation of fields occupied by diseased animals, become soiled by their dejections; the germs thus deposited may contaminate the litter and fodder and thus gain entrance to the alimentary canal of herbivora. It is not necessary in this case, as was at first believed, that the animals ingest at the same time

Dr. Trumbower. (*Rep. Board of Live Stock Commissioners of Ills.*, 1893.)

A few cases in cattle, in which the diagnosis was confirmed by competent bacteriologists are reported from Delaware. (*Fifth An. Rep. of the Del. College Agric. Experiment Sta.*, 1892, *p.* 45.)—D.]

substances capable of wounding the mucosa, thus opening the way for the bacteridia. It has been shown that the spores of charbon can be absorbed in the absence of all intestinal wounds. However, the non-sporulated bacilli are killed by the gastric juice so that their action can only take effect in the passages anterior to the stomach, and by a solution of continuity. Experiment has shown that the addition to contaminated fodder of substances capable of excoriating the mucosa (thistles, husks of barley) increases the mortality from charbon. At the wounds thus produced there is first developed a tumor, from which invasion proceeds by way of the lymphatics.

We have seen that, if the charbon bacilli have themselves great resistance, this faculty is possessed to a still greater degree by their spores. It is, morever, under this last form that the contagion of charbon persists on certain farms in such a way as to give the disease an endemic character. The contagion having once been deposited on a field, even on a limited part of the same, it remains there for years, contaminating the vegetation which grows upon it. The spores have been found on the surface of the ground above a grave closed for twelve years. The disease occurs during the pasturing season or in winter, according as the herbage is consumed while growing or after harvesting, in the form of hay. The most frequent cause of these endemic foci resides in the manner of burial of charbonous carcases: in spite of the layer, of greater or less thickness, by which they are covered, the spores produced in these bodies* find their way to the sur-

* [It seems to be well established that spores are not formed in

face. It is asserted that this migration is chiefly due to the agency of earth-worms. We know that these worms, in excavating the passages through which they wander, swallow a certain quantity of earth which they then expel in the form of tortuous rolls; we also know that they make regular nocturnal peregrinations to the surface of the soil in order to feed on the herbage. Now, M. Pasteur has found, in these little rolls of earth deposited above a grave containing a charbonous carcase, the spores or germ corpuscles of the disease. These spores do not necessarily always remain limited to the part of the field on which they have been deposited; they may be more or less widely disseminated by winds, and more especially by water. If we admit as established the possibility of the multiplication of the bacilli of charbon in swampy soil we can readily account for the persistence of the disease in certain localities.

The charbon bacilli may penetrate into the blood directly or indirectly; in this last case they first make their way into the tissues, multiply there, and excite an inflammatory engorgement, then gain the lymphatic glands and, finally, the blood. They then act by a complex mechanism. By reason of their aërobic character they rob the blood of part of its oxygen and thus give rise to asphyxia. This asphyxiating action, however, is contradicted by the investigations of M. Chauveau who has demonstrated the presence of a normal proportion of oxygen in the blood of a sheep

the blood or organs of buried cadavers (*Feser*, *Kitasato*, *Kitt*); they are found, however, on the surface of these bodies where they are soiled with blood, excretions, etc., and also in the digestive canal both before and after death.—D.]

in the last stage of charbon. It is especially by its secretions that the bacteridium acts upon the organism. The nature of these substances is as yet imperfectly understood, but their action is demonstrated by the following experiment: M. Chauveau transfused to a healthy sheep a large quantity of blood coming from a sheep affected with charbon, and, as a consequence, noticed the development of the general symptoms of the disease. This immediate result can only be explained as a chemical poisoning, an opinion which was, moreover, confirmed by the microscopic examination, which showed the disappearance of the bacilli from the transfused sheep. It is to this intoxication that we must ascribe the phenomena of nervous depression or temporary excitement, which are so marked in charbonous subjects.

Charbonous blood filtered through porcelain acts on the red corpuscles of normal blood, rendering them viscid and glutinous. This property is communicated to it by the substances secreted by the charbon bacilli and it explains the peculiar alteration of the blood in this disease.

Finally, the bacilli act mechanically; on account of their number and the abnormal viscidity of the blood, they form plugs in the interior of the capillary vessels, thus occasioning blood stases and superficial and deep hemorrhages. This is undoubtedly the cause of the final passage of the bacteridia into the milk, urine, and through the placenta; it is also the cause of the large and characteristic swelling of the spleen, as well as of the hemorrhages and general circulatory disturbance.

Attenuation. Vaccinations.—The virulence of charbon cultures is very stable; this is due to the presence of the spores which are little subject to change. When it is desired to attenuate these cultures it is necessary to begin by preventing the formation of spores. Pasteur succeeded in this by cultivating the charbon bacilli at the temperature of 42° to 43°. Multiplication of the bacilli still continues, but spores are no longer formed. Now, if such cultures are kept in contact with the air their virulence rapidly diminishes; after twelve days they no more kill adult guinea pigs, and vaccinate the rabbit and the sheep; the power of vegetation, however, still persists, and is absolutely extinguished only at the end of one and a half months, on an average. The bacteridium which has become asporogenous at the temperature of 42° to 43° then loses its virulent properties, retaining only those of an ordinary saprogenic microbe, and, finally, it loses all vitality.

The culture loses its pathogenic power little by little; it ceases to be fatal first for the large animals, then for small adults, and finally for small animals only a few days old. Now, each degree of virulence can be perpetuated separately by cultivating at 42° to 43° the different varieties obtained, each of these varieties transmitting its special virulence to its descendants. The degree of attenuation of each of these varieties can be preserved if the precaution be taken to frequently transfer to fresh culture media; but it is not possible to entirely avoid the attenuation which finally takes place. In order to definitely fix these varieties it suffices to return them to 37°; they then form spores possessing, in embryo, the special virulence of the bacilli from which they came and capable of transmitting this virulence to new generations of bacilli cultivated at 37°.

The least virulent varieties obtained produce a benignant disease which leaves behind it immunity for the varieties less attenuated, and thus are obtained vaccines of different degrees of intensity, which may be employed in succession. In practice two vaccines only are used, of which one or two drops are injected at an interval of ten days.

In the general part we have described the other methods of attenuation of the charbon bacillus—by compressed oxygen, by heat and by antiseptics—and the vaccination processes which have been derived from them.

*Symptomatic charbon.**

This disease, formerly confounded with charbon properly so called, has been separated from the latter as a result of the investigations of MM. Arloing, Cornevin, and Thomas. It is characterized, first, by the symptoms of a more or less intense fever and by the appearance of a specific tumor upon the body, neck, or upper part of the limbs. This tumor is almost constantly found in the muscular masses; it consists, at first, of a painful and progressive inflammatory engorgement, of firm and uniform consistence; it rapidly extends in area and in depth and, later, becomes insensible, crepitant, and resonant at its center (emphysemato-gangrenous tumor). The general symp-

* [Also occasionally referred to in this work as "bacterial charbon." Synonyms: Symptomatic anthrax, black-quarter, infectious emphysema; *Lat.* Sarcophysema hæmostaticum bovis; *Ger.* Rauschbrand.—D.]

toms become aggravated with the progressive evolution of the local lesion and the subject succumbs in thirty-six to forty-eight hours. When the disease has been somewhat prolonged, other tumors with the same characters not infrequently supervene. The distribution of these secondary tumors is not in relation with that of the lymphatics; their development appears to take place through the intermediation of the blood.

At the autopsy the local lesion is the predominant feature, the invaded muscles are friable and more or less darkened in color (charbon); their fibrous bundles are readily dissociated; the fibers retain their striation but their contents is broken up into hyaline, vitreous blocks (hyaline degeneration). The intra-muscular connective tissue, as well as that which surrounds the muscular masses, is thickened and infiltrated with a yellowish serosity; this œdema sometimes assumes considerable proportions. The formation of gas, inherent to the life of the germ, causes the detachment of the tissues, a true localized emphysema in the central part of the tumor. The gases produced are, chiefly, carbonic acid and marsh gas.

The lymphatic glands in relation to the tumor are reddened, ecchymosed and infiltrated.

After death the bodies very rapidly putrefy.

The animals which spontaneously contract the disease are cattle, sheep and goats. The receptivity of the first is not the same at all ages; calves of less than six months do not contract the natural disease, and cattle of over four or five years also seem to escape.

Microbe.—The pathogenic agents in this disease are straight rods, isolated or occasionally associated in pairs, measuring in the adult, non-sporulated condition 5μ to 8μ by 1μ; after fruiting they may attain larger dimensions—10μ by 1·3μ; as long as their contents are homogeneous they are cylindrical, but their form changes with the appearance of the spores; generally the spore is terminal and gives the bacillus a bell-clapper or club-shaped appearance; sometimes it is central, the bacillus then becoming spindle-shaped. The spore is solitary, ovoid, and very distinct; it occupies one-third of the length of the element.

The *bacillus Chauvœi* is endowed with oscillatory motion. It is strictly anaërobic; hence we should not expect to find it in the blood, at least during life; after death, when the oxygen is no more renewed by the pulmonary exchanges, it penetrates into this fluid.

Fig. 7.

Bacilli of symptomatic charbon, non-sporulated and sporulated. (M. and L.)

Action of physical and chemical agents. The virus withstands extreme cold and after being dried also resists for a considerable time the action of high temperatures; it is not destroyed by exposure for two hours to 80°, or for twenty minutes to 100°; on the other hand, it perishes in two minutes in boiling water.

The serosity dried at about 35° retains its virulence for more than two years.

Putrefaction has no effect on the bacillus of symptomatic charbon.

The bacilli, when they have escaped from the cadavers and been deposited upon the soil, preserve their virulence for a long time if the external conditions permit of their rapid desiccation; under other conditions they are more or less quickly attenuated and finally destroyed by the oxygen of the air. The same thing occurs when they have penetrated to a slight depth in permeable soil. But when they are buried at a sufficient depth, especially in a compact clayey soil, they are preserved for a considerable time on account of the absence of oxygen and light, and they may even, by virtue of their anaërobic faculty, propagate themselves there.

Strong antiseptics are fatal to the bacillus Chauvœi; Arloing, Cornevin, and Thomas have observed this toxic action after twenty-four hours' contact with sublimate, at 1 to 5,000, chlorine gas, and two per cent solution of carbolic acid; on the other hand, sulfurous acid, quick lime, and 90 per cent alcohol do not destroy its virulence.

Cultures.—Cultures of this microbe are very difficult to obtain; they are only possible when excluded from oxygen, in a vacuum, or in presence of an inert gas, such as carbonic acid.* Bouillons to which have been added glycerin and ferrous sulfate, or gelatin and sugar, are the media to be preferred. The fluid rapidly becomes turbid and the seat of an intense disengagement of gas; it exhales a pronounced odor of rancid butter.

Attempts at cultivation on solid media have also been successful. Upon gelatin there are produced spherical verrucose colonies, which fluidify the medium and give rise to a lively production of gas; cultures upon agar have a penetrating acid odor. Virulence does not long persist in generations obtained by

* ["Grows in an atmosphere of hydrogen but not in carbon dioxide." Sternberg: *Manual of Bacteriology*, p. 494.—D.]

artificial culture; it becomes obliterated in four or five transfers.

The material for inoculation of culture media may be taken from the juice obtained by scraping the center of a tumor, from the peritoneal serosity, and from the blood; but they only appear in this last fluid after death; as they are present in small numbers it is well to allow them to multiply by keeping a quantity of this blood in the incubator for twenty-four hours.

Research and coloration.—The bacillus Chauvœi exists in great abundance in the muscular tumor, suspended in the fluid with which it is infiltrated and which is interposed between the contractile elements. In the last moments of life and after death it is also met with in small numbers in the blood; finally, it exists in abundance in the bile and in the peritoneal serosity.

Simple methods of staining are alone successful; the various hydro-alcoholic solutions are available for this purpose, but Löffler's method should be preferred for sections.

Experimental inoculations.—The species to which it is possible to communicate the experimental disease are the ox, sheep, goat, and guinea pig. The rabbit is refractory. The ass and the horse only contract a local engorgement.

The receptivity of the various species for this bacillus sufficiently differentiates it from that of Pasteur's septicæmia which it much resembles in its physical characters, staining proclivities, and its anaërobic faculty, and in the emphysemato-gangrenous lesions which it occasions. The septic bacillus is pathogenic for all species except the ox; the bacillus of symptomatic charbon is pathogenic only for ruminants and the guinea pig.

Dermic inoculation with the lancet or by superficial scarifications is nearly always unsuccessful, whilst the introduction of the virus into the subcutaneous cellular tissue or into the muscular tissue gives positive results. However, it is necessary to take into account the dose injected and the place in which the injection is made. Very small doses do not produce the disease but confer immunity; similarly, a dose which would be fatal if injected in a favorable place is inoffensive when inoculated in the cellular tissue of the tail and of the extremity of the limbs. The experiments of M. Arloing have shown that this local immunity, which had been established by clinical observation, depends upon the greater density of the connective tissue and on the lower temperature of these regions. He succeeded in overcoming this immunity by heating the region or by lacerating its cellular tissue.

After the insertion of the virus in the subdermic tissue there results a painful and progressive inflammatory engorgement which extends to neighboring regions and in the guinea pig leads to death in twenty-four to forty-eight hours.

The intra-venous inoculation of small doses, but doses which would be fatal by the subcutaneous method (three to five drops of juice in young bovines, three-tenths of a drop in sheep), excites a rather intense febrile reaction; tumors are not produced unless some of the virus has been deposited in the cellular tissue surrounding the bloodvessel. The subject thus inoculated is vaccinated against the temporary reaction of a second intra-venous injection, and also against the subcutaneous inoculation of an otherwise mortal virus; the reaction of the latter then consists only in the formation of a curable abscess, the pus of which contains the virulent germs.

If the dose which can be tolerated in the vessels is exceeded, the typical disease ensues with the development of tumors. Similarly, if, after the intra-vascular injection of a vaccinating dose, a hemorrhage is produced in the connective tissue, a specific, fatal tumor appears in the place of the solution of vascular continuity, the bacilli having penetrated into their media of predilection.

The disease is transmissible by way of the respiratory passages, the result being the same as by the blood-vessels. Finally, the disease can be transmitted by way of the intact digestive canal if the virus is very active; the tumors then appear in places remote from the point of entry.

Etiology and pathogeny.—Symptomatic charbon is endemic in certain countries; it prevails especially during summer, its ravages being however less important than those of bacteridian charbon. Animals inoculate themselves accidentally, and as small doses confer immunity and this is transmitted from the mother to the fœtus, it results that part of the animals exposed to the contagion escape its fatal effects. But those which at first receive a sufficient dose quickly succumb.

The virus seems to be capable of entering by different ways. Wounds of the external integument in favorable regions are more especially suited to its evolution; but the inhalation of dust charged with dried virus, as well as the ingestion of forage soiled by very active virulent matters, can also occasion the disease. In the case of cutaneous wounds, the characteristic tumor develops at the place of inoculation itself; in the two latter contingencies the bacilli multiply in the blood and determine tumors in places where they meet with an opening by which they may penetrate into the connective or muscular tissue. Rupture of some of the fibers of the muscles, alteration of the vascular endothelium (perhaps by the products of the bacilli), the production of even a slight wound which may be overlooked, are all so many factors on which depend the seat of the primary and secondary tumors.

Attenuations. Preventive inoculations.—The virus becomes spontaneously attenuated when it is left in contact with the air; diminution of its virulence can be obtained artificially by means of antiseptics and heat. The latter agent supplied to MM. Arloing, Cornevin, and Thomas, the means of preparing virus of various degrees of intensity. The natural serosity of the specific lesions can be attenuated to different degrees by a temperature of from 65° to 70°, maintained for a greater or less length of time. These authors, however, operated by preference with serosity dried at the temperature of 30° to 35°; the dried virus is, in reality, more fixed than the fluid serosity, because the spores are much more resistant when dry than when in a moist condition. This dried virulent substance withstands temperatures of 80° to 90° without losing any of its activity; reduced to powder and moistened, then brought to temperatures varying between 60° and 110°, it becomes progressively attenuated. The virus, attenuated to such an extent that it is no more fatal, vaccinates the organism against more active virus. The French authors recommend the successive employment of two vaccines, one exposed to 100° the other to 90°, during seven hours. These two vaccines are dry when taken from the oven in which they have been prepared; the dose employed is one centigram of the powder diluted in a gram of water for each animal. The vaccine prepared at 100° is first used, and then, after eight days, the other. The inoculations are made in the cellular tissue of the ear, or in the internal face of the end of the tail. The autumn or the end of winter are the seasons selected for the operation. It is necessary to guard against inoculation during very hot weather as an elevated temperature increases the activity of the virus.

The consequences of the vaccinal inoculation are local and general. At the inoculated point an engorgement develops, generally of small extent and healing spontaneously. At the same time there is observed a febrile reaction, which indicates the existence of the disease in a mild form.

Kitt recommends a single vaccine prepared at 90°.

When the virus has been thus artificially attenuated it can regain its original activity by successive passages through the bodies of young guinea pigs, and also under the influence of lactic acid, etc., as already described.

Animals can also be vaccinated by injection of the natural virus either in the cellular tissue or into the blood. We have seen above, that inoculation of small doses in the connective tissue confers immunity in place of the disease, and that the blood tolerates comparatively large doses of the virus. Injection into the veins, however, must be made with the greatest care in order to avoid the accidental deposition of any of the virus in the surrounding tissue, hence, in practice, the attenuated viruses are generally preferred to the natural viruses, for the production of immunity.

*Rouget of the pig.**

This disease, peculiar to the pig, is infectious and contagious; it chiefly attacks adult animals and those of the improved breeds. It manifests itself by a very intense febrile reaction, by red or purple patches, at first discreet and afterward confluent, upon the integument, by a diarrhœa more or less intense, succeeding to constipation, and often by cough. The redness of the skin may be absent in very acute cases.

The duration of the disease is always short, on an average, two days; it may, however, last four or five days, and, on the other hand, may occasionally be almost fulminating in character.

Death is the usual termination, but a considerable proportion of pigs may recover. Moreover, all the the pigs which have been in contact with diseased animals do not necessarily contract the disease.

The autopsy discloses a general congested condition of the capillaries. There is injection and serous infiltration of the skin and subcutaneous cellular tissue, injection with petechiæ and sero-fibrinous exudation of the peritoneum, pleura and pericardium, changes of the same kind in the gastro-intestinal canal in which the mucosa is reddened, thickened and infil-

* [*Eng.* Swine erysipelas; *Ger.* Rothlauf. The disease has not been recorded in the United States or Canada.—D.]

trated; in many places the epithelium is desquamated, and occasionally there are ulcerations in way of formation.

The blood vascular glands (spleen, lymph nodes, Peyer's patches) are tumefied by congestion, exudation and extravasation; the kidneys, lungs and heart always show injection and even extravasations. In rare cases there are multiple lesions of broncho-pneumonia.

The germ of rouget is a very fine, cylindrical bacillus recalling that of mouse septicæmia; (1) it measures 1μ to 2μ in length by 0·1μ to 0·15μ in thickness. It is found in the blood, especially in the fine capillaries, in contact with their internal wall; it also exists in the exudates and in all the diseased organs: liver, spleen, kidneys, lymphatic glands, marrow of bones, etc.; in the fæcal matters, and in the urine. This agent should not be confounded with a large rod which is found in the blood in most diseased pigs and which does not

(1) Mouse septicæmia was obtained experimentally by Koch, by inoculating putrefied blood under the skin of these animals. The disease is easily transmitted to house mice, whilst field mice are refractory. The microbes show much resemblance with those of rouget; they have the same form, the same dimensions, are rather anaërobic than aërobic, and both take the Gram and Weigert stains. In gelatin, stab cultures of mouse septicæmia give colonies with radiating branches like those of rouget, but the rays are confounded with each other, whilst those of rouget are distinct; gelatin is not fluidified by either. Potato is poorly or not at all adapted to their vegetation. Both become much elongated in cultures. The pigeon succumbs with equal rapidity to rouget and mouse septicæmia. The rabbit, on the other hand, is much more sensitive to the former of these diseases. Both produce, in the pigeon and the mouse, enlargement of the spleen and congestive lesions of the different organs.

show any specific pathogenic property. This last germ is never seen in the experimental disease and probably comes from a secondary invasion occurring in the diseased animal. The point of departure of this collateral infection appears to be the intestine, in which the bacillus just mentioned is found in abundance; it is present in the blood in smaller numbers the further removed the latter is from the abdominal cavity.

Pasteur and Thuillier, who first described the bacillus of rouget, describe it as of a figure 8 form, but this was an error of observation which our staining methods and improved instruments have corrected.

The bacillus of rouget is non-motile.

Cultures.—The bacillus of rouget is especially anaërobic but it also grows in contact with the air; it multiplies at a temperature as low as 20°, but grows especially well in the incubator.

The blood and the pulp of the various diseased organs may be used for the inoculation of culture media. This material will more likely be pure when taken from a part at some distance from the abdominal cavity.

The germ grows well on the various culture media: on potato, especially in presence of oxygen, the growth is feeble.

In bouillon it produces, after forty-eight hours, a slight uniform turbidity, which afterward becomes deposited as a whitish gray sediment.

Stab cultures in gelatin are characteristic. The germ multiplies especially in the deeper parts and forms along the course of the puncture a track from which radiate silky tufts, giving to the whole the aspect of a bottle brush. This characteristic form is only obtained with gelatin of firm consistence. The bacillus of rouget appreciably lengthens in its cultures.

Research and coloration.—The specific germ can be demonstrated in the blood, in the fluid exudates, and in sections of the various tissues affected; on account of its small size it is advisable to examine only stained preparations and to employ a magnification of 800 diameters, at least. The bacillus of rouget is stained by the hydro-alcoholic solutions of all the aniline colors; the color is not removed by the reactions of Gram and Weigert.

Fig. 8.

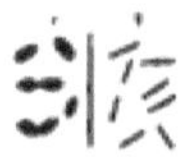

Bacilli: 1, of pneumo-enteritis of the pig; 2, of rouget.—(M. and L.)

Experimental inoculations.—The bacilli of rouget are fatal for the pig, rabbit, mouse and pigeon. The three last species are especially sensitive and succumb to subcutaneous, intra-peritoneal, or intra-venous inoculations of a virulent product, whether natural or derived from culture.

Subcutaneous inoculation may fail in the pig; ingestion, although generally communicating the disease, is also uncertain in its results. The mouse and the pigeon succumb in two to four days; the pigeon takes the form of a ball of feathers, as in cholera; the rabbit dies in three to six days. The lesions noticed at the autopsy consist of a hypertrophy of the spleen with congestion of the liver and lymphatic glands.

The guinea pig, rat, dog, and chicken are refractory to experimental inoculation.

Etiology and pathogeny.—Although some tests of infection by the digestive passage performed in Germany by Lydtin were unsuccessful, there is reason to believe that, under natural conditions, the virus gains entrance by the digestive canal.

The virulent matter ingested by healthy pigs will consist especially of the intestinal dejections of the diseased. Lack of proper attention to the feeding troughs, and defective conditions of the pens (ventilation, lighting, cleanliness) are of a nature to favor the propagation of rouget, as well as of other diseases of the pig.

Adult hogs, especially those of the improved breeds (English), are most liable to contract the disease; young pigs and those of native breed are more resistant.

Attenuation. Vaccination.—Rouget of the pig becomes well acclimated in the pigeon and the rabbit, and in these two species acquires great virulence; but whilst its repeated passage through the organism of the pigeon renders it more active for the pig, passage through the rabbit, on the contrary, diminishes its virulence for the pig. This attenuation, after a certain time, is such that the virus coming from the rabbit no more kills the pig; it, however, makes it sick and confers immunity on it against strong virus. The attenuation thus obtained persists in cultures afterward made in ordinary bouillon and these cultures can be used to vaccinate the pig.

In practice two vaccines only are employed, and these in succession at ten days interval; a feeble vaccine is used first and then a second, the virulence of which is much stronger.

Young pigs being much less susceptible to the disease and therefore to the virus, the period of youth should be preferred for vaccination. The immunity obtained lasts only one year, but this term is sufficient for the needs of breeding and fattening.*

Pneumo-enteritis of the pig, hog cholera.

Pneumo-enteritis of hogs is an infectious and contagious disease which was long confounded with rouget. It was described, in the first place, in America, by Salmon under the name of *hog cholera;* in France it has been investigated by Riestsch and Jobert, and Cornil and Chantemesse, in connection with the epizootics at Marseille and at Gentilly; it was studied by Selander in the swine of Sweden and Denmark.

This disease chiefly attacks young animals and is nearly always fatal. It manifests itself by symptoms the description of which differs somewhat in the different countries in which it has been observed.

According to Salmon, the disease may be acute or chronic. In the latter case inappetence is observed along with persistent diarrhœa and slow emaciation of the diseased animals. When the disease is acute the diarrhœa is more intense and sanguinolent. In both cases the intestine is much altered, principally the large intestine. The latter presents ulcerations and considerable thickening of its mucosa when the disease has been slow; when the evolution has been rapid the lesions assume a hemorrhagic character and affect not only the cæcum and large colon, which are

* [Lorenz has introduced a method of protective inoculation by the use of the blood-serum of swine which have previously been immunised against rouget. (*Centralblatt für Bacteriologie*, xiii—11, 12.)—D.]

much injected and ulcerated, but also the spleen, liver, kidneys, and mesenteric glands. Generally the lungs are unaffected; nevertheless, some foci of hepatization may be noticed in the last stage of the slow form of the disease. The skin of the neck and abdomen, and sometimes of the whole body, is reddened.

Intestinal lesions, therefore, predominate in the disease studied by Salmon.

The disease studied in France is characterized by an intense fever with considerable prostration of the affected animals, staggering gait and sometimes paralysis. An intense and fetid serous diarrhœa soon supervenes; this is often preceded, and occasionally followed, by constipation; at the same time, or a little later, symptoms of pulmonary trouble become evident: fitful hoarse cough, accelerated and embarrassed respiration, and abundant mucous discharge from the nostrils.

The most conspicuous symptoms vary according as the intestinal or pulmonary troubles predominate. In the epizootic at Marseille enteritis was constant and the pulmonary lesions incidental; on the contrary, broncho-pneumonia was the dominant feature in the epizootic at Gentilly. MM. Cornil and Chantemesse think that this peculiarity depends upon the mode of entrance of the virus. The Gentilly hogs were contaminated at the abattoir by inspiring air charged with virulent dust, whilst those of Marseille contracted the disease by ingestion of contaminated food.

In the course of the disease a diffuse inflammation often develops on the skin of the lower part of the abdomen, on the perineum, groins, limbs, and at the root of the ears, these regions then taking a more or less pronounced red or violet color, and thus tending to increase the chances of confounding it with rouget. These cutaneous changes, however, are less constant than in the last disease.

Pneumo-enteritis is of rather long duration: twenty to twenty-five days on an average, never less than eight to ten days; it may extend to five or six weeks; it is very contagious, and few hogs which have been exposed to the contagion escape.

In very rapid cases the autopsy shows, beside ecchymoses disseminated through the connective and inter-muscular tissue, peritoneum, pleura, pericardium, and heart, a violent inflammation of the stomach and intestines, with interstitial hemorrhages and erosions or ulcerations at Peyer's patches; the mesenteric glands are voluminous and infiltrated; the lungs are normal or show lobular foci of hemorrhagic congestion. When the evolution of the disease has been slow the lesions are better defined; those of the intestine, cæcum, and large colon are especially remarkable; the wall of these organs is considerably thickened and indurated, and has become rigid. The swelling and induration chiefly affect the Peyer's patches; these are the seat of a necrotic process which leads to the formation of grayish colored diphtheritic exudates and ulcerations of greater or less extent, both in area and in depth. The inflammation sometimes extends to the peritoneum.

The lungs show lesions of broncho-pneumonia at a more or less advanced stage; pleurisy is also occasionally present.

The mesenteric and bronchial lymphatic glands are tumefied and sometimes partly caseous.

Microbe.—The germ of pneumo-enteritis belongs to the group of bacteria showing a clear central space, that is, those which take the stain better at their margins than in the center. It is ovoid, measures 1μ to 2μ in length by 0·4μ to 0·6μ in thickness. It is motile, aërobic and facultatively anaërobic. It does not form spores.

Fig. 9.

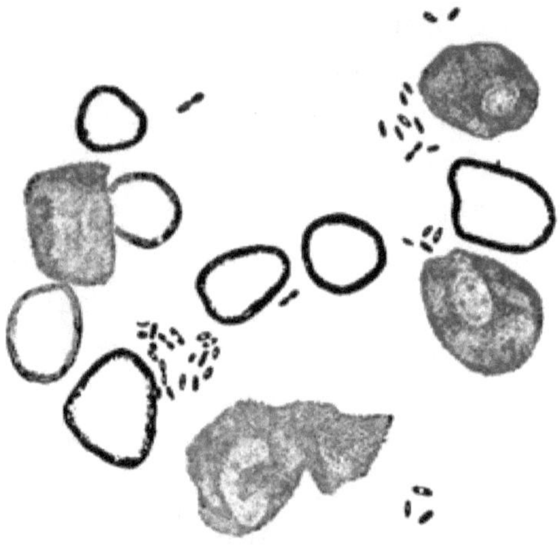

Hog cholera bacilli in spleen of guinea pig; cover-glass preparation. ×1200.—D.

Action of physical and chemical agents.—The bacillus of pneumo-enteritis is destroyed by a temperature of 58°, maintained during from fifteen or twenty minutes. It preserves its vitality in spite of desiccation for nearly two months. It vegetates and multiplies in water at the ordinary temperature of summer; it retains its vitality for more than fifteen days in sterilized water.

Those authors who have studied the action of chemical agents on this microbe especially recommend for its destruction mineral acids and sulfate of copper. MM. Cornil and Chantemesse recommend the following solution for the disinfection of pens and other infected objects:

Water	100
Carbolic acid	4
Hydrochloric acid	2

Cultures—Cultures on artificial media succeed well at temperatures varying from 18° to 45°.

Bouillon becomes turbid without showing any special characters.

Gelatin, inoculated in superficial lines, shows a raised growth of white or bluish-white appearance, with irregular, often lace-like, borders; the medium is not fluidified. Inoculated by puncture the culture shows itself in the form of rounded colonies covered with crystalline projections.

The culture upon potato is remarkable for its clear brown color, which gradually becomes deeper with age.

Sowings ought to be taken from the parenchymatous organs: liver, lungs, lymphatic glands, or from the blood.

Schweinitz has isolated from cultures a toxic ptomaine (sucholo-toxin) and a special albumin (sucholo-albumin).

Research and coloration.—Salmon's bacillus readily takes up the different aniline colors; simple methods of staining are alone applicable: hydro-alcoholic solutions, Löffler's blue, etc. The methods of Gram and Weigert, and Kühne's violet completely fail.

Experimental inoculations.—The disease is inoculable to the mouse, rabbit and guinea pig. The pigeon takes it only from large doses.

In the *mouse* the microbe assumes larger proportions than in the pig, and multiplies abundantly.

The *rabbit*, inoculated under the skin, succumbs in

from three to eight days according to the dose inoculated; the lungs are gorged with blood, the intestines are frequently the seat of a violent inflammation with resulting diarrhœa; the spleen is tumefied and, along with the liver, often shows white necrosed foci. At the place of inoculation a creamy mass is found, also the product of a coagulative necrosis.

The *guinea pig* dies in the same time, and with the same lesions, as the rabbit.*

The *pigeon* is very resistant to pneumo-enteritis. It withstands small doses inoculated in the pectoral muscle although the inoculated point becomes the seat of a sequestrum similar to that produced by chicken cholera. Very large doses kill it in less than two days.

The *chicken* is refractory.

The *pig* is difficult to contaminate by hypodermic injection; on the other hand, it succumbs ninety times out of a hundred to ingestion of virulent products. It contracts the same disease by inhalation of dust containing virulent matters held in suspension in the air, and also by intra-vascular injection.

Etiology and pathogeny.—The cause of pneumo-enteritis resides in the bacillus discovered by Salmon; the disease generally shows itself on a farm in consequence of the introduction of an infected hog; it is transmitted from one hog to another by the ingestion of food or drink soiled by the intestinal dejections and nasal discharge of the diseased animals.

* [The guinea pig appears to be somewhat less susceptible than the rabbit, and rarely shows the characteristic necrotic foci found in the liver in the latter animal.—D.]

MM. Cornil and Chantemesse also conclude that the germ may penetrate by the respiratory passages and according to the way by which it is introduced, the disease will act more particularly upon the intestines or upon the lungs.

According to Salmon, the natural virus is subject to considerable variations of activity, and this explains the sometimes rapid and sometimes slow course of the disease.

The local lesions which follow its penetration assume the general character of an inflammation with a tendency to early mortification; this tendency is sufficiently demonstrated by the diphtheritic exudates and ulcerations of the intestinal mucous membrane and the coagulative necrosis, under the form of caseous masses, in the glands of diseased hogs. This character also shows itself in the mouse, rabbit, guinea pig and pigeon which have been subjected to experimental inoculation.

The local alterations lead to emaciation of the affected animals, but this enfeeblement is complicated with an intoxication. We have already seen that a toxic ptomaine has been isolated from cultures of pneumo-enteritis, and it may be that this poison, secreted in greater abundance by very virulent germs, is the cause of the vascular changes met with in acute cases.

The resistance of the bacteria to desiccation, and the facility with which they multiply in water at the ordinary temperature, are conditions which favor the persistence of the disease in one place and the production of new centers of infection.

Attenuation. Vaccination.—Attempts at attenuation

of the bacillus of pneumo-enteritis have been made by MM. Cornil and Chantemesse. They had recourse to the action of heat upon cultures. A culture maintained at 43° during seventy-four days produces only a local abscess in the rabbit but is still regularly toxic for the guinea pig. After ninety days the virus no more kills guinea pigs; they contract an abscess at the point of inoculation, whilst rabbits often escape even this lesion. The virus thus attenuated transmits its special virulence to its descendants and confers, on guinea pigs and rabbits which have received it, immunity for virus which has been heated only seventy-four days; the latter acts in the same way toward more active and the natural virus. It is therefore possible to vaccinate the rabbit and the guinea pig against pneumo-enteritis. Unfortunately this method of prevention, applied to the pig, has not given the same result, and a process of vaccination is yet to be found for this species.

The vaccine of rouget does not vaccinate hogs against cholera, a circumstance which increases the importance of the differential diagnosis.

Schweinitz has succeeded in vaccinating the guinea pig by means of soluble substances which he has isolated from cultures.*

* [Billings obtained protection against this disease by inoculation of pigs with cultures derived from mild cases of the natural disease. A certain proportion of the animals die from the inoculation. *Neb. Ag. Exper. Sta., Vol. 2, No. 4.* Selander (confirmed by Metschnikoff) found that cultures of the Danish swine-pest, increased in virulence by passage through pigeons, produced in the blood of rabbits very active toxines. When such blood was sterilized at 57° C. and injected in rabbits it conferred a solid immunity. The serum of rabbits thus vaccinated was also found to

Pneumo-enteritis of the sheep.—M. Galtier has studied a disease in the sheep to which, from its principal lesions, he has given the name of pneumo-enteritis, and which, according to this author, is caused by the germ of the disease of the pig which has just been described. This affection sometimes occurs in an epizootic form in sheep, and may make great ravages in affected flocks. In several cases the disease had originated in consequence of the introduction into the sheep-folds of swine recently purchased and which had contracted pneumo-enteritis in the market pens. Once established in sheep, it transmits itself with great facility from sheep to sheep.

The general symptoms consist in lassitude and general loss of vigor of the affected animals, with inappetence, loss of rumination and the appearance of a more or less intense fever. These symptoms are soon succeeded by bloating, fetid and exhausting diarrhœa, cough, accelerated respiration, mucous discharge from the nostrils sometimes streaked with blood, and the special symptoms of a broncho-pneumonia or of a broncho-pleuro-pneumonia. The skin and the visible mucous membranes take a more or less vivid red color, sometimes mixed with hemorrhagic points. In pregnant females abortion is often observed although the mother does not necessarily succumb to the attacks of the disease.

The disease may show various degrees of intensity; it is sometimes very severe and kills in a few hours, or days, sometimes benign and passes unperceived. Con-

possess immunising properties. *Annales de l'Inst. Pasteur*, 1890, *p.* 546. These results are not obtained with the American disease. (Smith & Moore.)—D.]

valescence from the severe forms is always prolonged. The receptivity of sheep diminishes with age; thus, the disease is noticed to be more severe and more frequently fatal in young animals.

The bodies rapidly putrefy; the subcutaneous and intermuscular connective tissue is dotted with hemorrhagic points, sometimes with gelatinous exudates. The peritoneum, pleura, and sometimes the pericardium may be the seat of fibrinous inflammations. The mucous membrane of the fourth stomach, small and large intestine, is congested; it shows extravasated points and sometimes erosions; Peyer's patches are tumefied. The liver is also hyperæmic and dotted with petechiæ; sometimes it contains abscesses when the disease has been somewhat prolonged. There are disseminated lesions of broncho-pneumonia with infiltration and thickening of the interlobular connective tissue septa. The mucous membrane of the bronchi and of the trachea is reddened and thickened, and secretes an abnormal quantity of mucus. When the disease has developed slowly, caseous foci are not infrequently found in the lungs. The lymphatic glands of the mesentery and of the root of the lung are enlarged, congested and infiltrated.

The micro-organism which produces this disease of the sheep is identical, according to M. Galtier, with that of pneumo-enteritis of the pig. This author claims to have succeeded in transmitting this last disease to the sheep by inoculation, as well as in transmitting the disease of the sheep to the rabbit, guinea pig, dog, pig, goat, calf, to solipeds, to the chicken and to the pigeon. In the goat it resembles the pleuro-pneumonia (bou-frida) peculiar to this species; in solipeds it gives rise to symptoms recalling the typhoid disease; finally, it would explain epizootic abortion when it is spontaneously transmitted to cows.

Transmission is readily procured by hypodermic, intravenous and intra-pulmonary injection, less easily by the digestive canal. The virus loses its virulence by multiple transfers on artificial media and passages through individuals in which the disease develops slowly. On the contrary, its virulence increases in organisms very susceptible to its influence. Thus, according to M. Galtier, the pneumo-enteritis of the pigs at Gentilly is transmissible to sheep (contrary to the assertion of M. Nocard) when the substance inoculated is taken not from a culture but from a diseased animal.

The natural contagion occurs by ingestion, and especially by inhalation, of virulent products. The disease is also transmitted from the mother to the fœtus.

According to the researches of M. Galtier, which we have just briefly reviewed, pneumo-enteritis, which is generally considered to be peculiar to the pig, extends to all farm animals, especially to the sheep, bovines and solipeds. The disease being transmitted to the fœtus, calves coming from diseased cows which are or have been subject to coughing are born with the germ of the disease in them and die in a few days with the lesions of broncho-pneumonia and enteritis (pneumo-enteritis of calves).

Infectious pneumonia of the pig, swine plague.

This disease of swine has been described in Germany under the name Schweineseuche, and in America under that of Swine-plague. It is an infectious and epizootic disease characterized by the predominance of the pulmonary lesions. These consist of broncho-pneumonic foci with or without complications of pleurisy, enteritis, etc. The lung often contains caseous masses; sometimes it is gangrenous. The evolution is acute or chronic; in the first case the duration of the disease is from three to nine hours on an average.

There is present, in the pulmonary lesions, pleura, peritoneum and pericardium, a non-motile microbe, ovoid, from 1μ to $1{\cdot}2\mu$ long by 0.6μ in thickness, staining only at its extremities.

The disease is inoculable to the mouse, rabbit, guinea pig, chicken and pigeon, but large doses are required to produce fatal results in the last three species.

It is transmitted from one hog to another by inhalation, probably also by ingestion, and perhaps by accidental cutaneous inoculation.

*Differential diagnosis of rouget, pneumo-enteritis, and swine-plague (peste-porcine.)**

ROUGET.	PNEUMO-ENTERITIS.	SWINE PLAGUE.
Course rapid: 2 to 5 days.	Course slow; 20 to 25, days, never less than 8 to 10 days.	*
Attacks especially adult hogs.	Attacks especially young pigs.	
General symptoms predominating.	Gastro-pulmonary symptoms predominating.	Pulmonary localiz tion predominating.
Redness more extended, more constant.	Redness less extended, less constant.	
Congestive lesions of all the organs, with petechial extravasations and exudative inflammations of serous membranes.	Inflammatory lesions of a necrotic character in the intestine posterior to the ileo-cœcal valve, and in the mesenteric and bronchial glands.	Broncho-pneumonia with caseous foci.
In very slow cases the intestine may contain ulcerations at Peyer's patches as in pneumo-enteritis.	When the disease is rapid the intestinal changes may not have reached their usual stage and thus may cause the disease to be mistaken for rouget. This mistake is more easily made in cases in which the pulmonary lesions have not had time to develop.	
Bacilli cylindrical, non-motile, stained by the Gram and Weigert methods.	Bacteria ovoid, motile, not stained by the Gram or Weigert methods. The center stains much less than the periphery.	Bacteria ovoid, non-motile, staining only at their extremities.
Especially anaërobic.	Especially aërobic.	
Inoculable to the mouse, rabbit and pigeon, but not to the guinea pig.	Inoculable to the mouse, rabbit, and guinea pig, to the pigeon only when very large doses are employed; not to the chicken.	Inoculable to the mouse and to the rabbit, and when large doses are employed to the guinea pig, pigeon and chicken.
Culture in gelatin has the appearance of a test-tube brush.	Cultnre has the form of globules covered with crystalline asperities.	

* [From the lack of distinctive names for the various infectious diseases of swine which have been described in different countries, the morphological and biological similarity of their germs, and the absence of any post-mortem lesions which are absolutely characteristic, there still exists considerable confusion as to their

Tuberculosis.

Tuberculosis is a disease too well known to require a description of its symptoms and lesions in this place. We will consider it only from the special point of view of its bacteriology.

The disease produces its greatest ravages in the human species, and then, with a frequency decreasing in the order in which they are named, it attacks the bovine species, the pig, horse, dog, and cat. Fowls also are decimated by this terrible scourge. The question as to the identity of human and avian tuberculosis has been much discussed and is still in dispute.

The distribution of the specific alterations is somewhat different in the different species.

In cattle they are most frequently found in the lungs, pleura, and thoracic glands, but are also common in the intestines, peritoneum, liver, kidneys, mammæ, and the corresponding lymphatic glands; they have also been met with in the meninges, inter-

identity. The names given in each of the lists below seem to refer to the same disease:

I. Rouget, in France; rothlauf, in Germany; swine erysipelas, in England (cases described by McFadyean, Jour. Comp. Path., Vol. IV, Part 4).

II. Pneumo-enteritis (Klein), swine fever, in England; svinpest (Selander), in Denmark; hog cholera (Salmon and Smith), swine plague (Billings), in America.

III. Schweineseuche, in Germany; swine plague (Smith), in America (uncertain).

The germ of the French disease of hogs described as pneumo-enteritis, as well as that of Galtier, has not been identified with that of either of the diseases included under lists II and III.—D.]

muscular connective tissue, bones, bone marrow, and the articulations. The disease is especially frequent in adult animals; when it exists in the calf it is located nearly always in the abdominal viscera, and more especially in the liver when the affection is congenital; but it can quickly extend to the thoracic organs. We have had the opportunity of establishing this extension to an excessive degree in the viscera coming from the abattoir of Brussels, the lungs and bronchial glands showing a generalization of miliary tubercles. In two cases of intra-uterine infection observed by MM. Malvoz and Brouwier the lung was exempt from lesions, these occupying the liver, hepatic, and bronchial glands.

The pig, although very susceptible to the experimental disease, appears to be rarely affected with tuberculosis; here it is the pulmonary form that predominates. M. Moule has communicated a case in which he observed extension to the pleura, ribs, and muscles. Microscopical and bacteriological researches have enabled us to connect with tuberculosis those scrofulous alterations of the cervical glands which are occasionally observed in the pig. According to M. Nocard, the disease in this species often develops with great rapidity and passes unperceived; in the chronic forms the bacilli are rare and seem to have lost part of their virulence; inoculated to guinea pigs they produce a disease of slow course; but the period of incubation becomes shortened again when these bacilli are inoculated from the first guinea pig to others in series. This property belongs also to the lesions of human scrofula.

In the horse two forms of tuberculosis exist; the

abdominal form, the most frequent, is characterized by confluent lesions in the spleen, mesenteric glands, liver, and intestines; in the thoracic form the changes occur chiefly in the lungs; the latter also develop, but more slowly, as a sequel to abdominal tuberculosis. According to M. Nocard, who gave us our first information upon tuberculosis of the horse, this disease is often accompanied with a polyuria of remarkable intensity.

In the dog the disease also occurs under the two forms observed in the horse. A considerable number of cases of tuberculosis in this species have already been communicated. We ourselves have seen two cases. In the first the liver was enlarged and infiltrated with a neoplastic tissue of a grayish color, which, under the microscope, was resolved into miliary tubercles destitute of giant cells. The hepatic and mesenteric glands were much tumefied, and partly caseous. The lung contained two nodules of the size of a pea. The second dog showed generalized tuberculosis of both lungs, and of the bronchial glands.

The ape is very susceptible to the disease, and very readily contracts it in our climate.

In fowls the abdominal viscera are most affected, and this often to an excessive degree. In this species the liver seems to be the place of predilection for the tubercles; sometimes the latter are absent, only a considerable enlargement with degeneration of this organ being observed, but the spleen, intestines, and peritoneum may, at the same time, be studded with lesions. The lung is rarely affected.

Microbe.—The efficient cause of tuberculosis resides in the bacillus of Koch. This shows itself under the form of a homogeneous rod, or more frequently composed of granules arranged in linear series. It is straight or bent in the form of an arc, sometimes S-shaped; it measures 2μ to 6μ in length by $0{\cdot}3\mu$ to $0{\cdot}5\mu$ in thickness. The rod is uniform in size throughout its extent.

According to Cornil and Babès the granular appearance is especially observed in bacilli long abandoned to the air; these authors have demonstrated the same granules in bacilli coming from cultures and they regard them as spores.

Fig. 10.

Tubercle bacilli. (M. and L.)

A certain number of the bacilli are observed to become considerably elongated and to swell at one of their extremities; we have often seen these abnormal forms in bouillon cultures of avian tuberculosis.

Action of physical and chemical agents.—Tubercle bacilli are killed by a temperature of 70°, maintained during ten minutes (Yersin). According to M. Galtier, heating at 71° during ten minutes does not suffice to sterilize tuberculous matter. Moist heat at 100° sterilizes it with certainty in a few minutes, but this is not the case when the virulent substance is in the dry state, for the dried spores withstand 100°.

The tubercle bacillus withstands freezing, putrefaction and desiccation. This last operation, when it occurs at temperatures near to 30°, is, indeed, an important means of preserving it. The bacillus retains its vitality for a considerable time in sterilized water (seventy days).

According to Yersin, carbolic acid, at 5 per cent, kills the bacillus in half a minute; sublimate, at 1 to

1,000, in ten minutes; absolute alcohol in five minutes; but contact with these agents must be much more prolonged in the case of tuberculous substances, sputa, etc. Hence, for the destruction of the virulence of these substances, moist heat should be preferred.

The tubercle bacillus resists the action of the gastric juice.

Cultures.—Vegetation of the bacillus on artificial media is not easily obtained; the operation, however, is only delicate for the first generation. The culture media most favorable for this bacillus are those which contain an addition of peptone, glycerin, and even of glucose, in proper proportions. The incubating oven should be kept at a temperature in the neighborhood of 39°, that temperature being best suited for the growth of the bacillus; multiplication is impeded by slight deviation from this temperature, and ceases altogether at about 35°. As the germ is aërobic, the culture medium should also be freely exposed to the air. The sowing should be taken from virulent products from young tuberculous animals in which the evolution is rapid, and should be implanted at once upon serum; it is transferred several times upon this medium before trusting it to others.

On serum, at the end of twelve to fifteen days, small, round, whitish grains appear, and slowly increase in size; these grains are slightly raised, dry, and scaly in appearance; their growth is very limited in the first cultures and they become confluent only after the fourth or fifth generation; vegetation is then more rapid and the whole surface of the serum becomes covered with a thin, dry film, studded with verrucose prominences.

Sowing of tuberculosis of mammals directly upon agar failed of results in the hands of MM. Straus and Gamaleïa, and even transference of serum cultures on to agar only succeeded well after four or five passages on the serum.

The bacillus, therefore, requires to become acclimated on an artificial medium in order to obtain a vigorous growth. If it is then transferred to animals it vegetates satisfactorily.

The appearance of cultures on agar resembles that of serum cultures.

In the case of avian tuberculosis much richer cultures are obtained directly upon serum. These cultures begin by rounded, whitish spots, waxy and moist, which after a few transfers produce a continuous layer of the same appearance, thus contrasting with the meager and dry film of human tuberculosis (Straus and Gamaleïa). Cultures on agar and on bouillon are also more readily obtained and more abundant than in the case of human tuberculosis.

In bouillons, after a few days, small flakes appear which gradually increase in size and fall to the bottom of the liquid without becoming dissociated; they break up into finer particles only when the vessel is shaken.

The tubercle bacillus also vegetates on potato, although this is not a very favorable medium for its growth.

Research and coloration.—Koch's bacillus fixes coloring matters with difficulty and hence requires prolonged exposure. In order to abbreviate the operation recourse is often had to heat. The coloration, however, if slowly obtained, is persistent even against the action of strong acids—nitric and sulphuric; upon this property are based the different methods of double staining which we will now describe. This property is also possessed by the bacillus of leprosy.

1. *Ehrlich's method.*

The staining fluid is composed as follows:

Aniline water (1)	9 cub. cent.
Absolute alcohol	1 cub. cent.
Concentrated alcoholic solution of fuchsin, methyl violet, or gentian violet	1 cub. cent.

Cover glasses remain in this solution half an hour at least, sections twenty-four hours.

Decoloration is obtained by a dilution of nitric acid—1 part nitric acid in 5 parts distilled water or in 10 parts alcohol. Cover glasses or sections should remain in the decolorizing fluid only from half a minute to a minute. They are then washed in distilled water and mounted. If it is desired to stain the background of the specimen so as to render the color of the bacilli more distinct, the preparations, after coming from the distilled water, are placed in a hydro-alcoholic solution of methylene blue if fuchsin has been the first stain, or in eosin, safranin, bismark brown, etc., if the bacilli have been stained violet; then they are mounted.

The time required by this method may be abridged by raising the temperature of the staining fluid to

(1) Aniline water is made in the following manner: A drop of aniline oil is added to a few cubic centimeters of distilled water in a test tube; the tube is vigorously shaken in order to dissolve the oil, and a few drops of absolute alcohol added to complete this solution.

near the boiling point before the introduction of the specimens to be stained. The necessary staining is thus obtained in a few minutes.

2. *Lubimoff's method.*

The staining fluid is composed of:

Water	20 cub. cent.
Boric acid . . .	0·5 grams.
Absolute alcohol . .	15 cub. cent.
Fuchsin	0·5 grams.

This fluid keeps indefinitely.

To stain the bacilli upon the cover glass a few drops of the solution are deposited upon its surface and heated for two minutes over a spirit lamp; the preparation is then rapidly decolorized in sulfuric acid diluted to 1 to 5, then rinsed with alcohol and immersed in a concentrated alcoholic solution of methylene blue; it is then washed with water, dried and mounted in balsam.

Sections of tubercular tissue are left during one or two minutes in the staining fluid, previously heated to near the boiling point, then passed for a few seconds into alcohol and thence, for one to two minutes, into a 1 to 5 dilution of sulfuric acid, again into alcohol and, finally, for one minute, into a hydro-alcoholic solution of methylene blue. Dehydration is accomplished by passing the sections through absolute alcohol and xylol and they are then mounted in balsam.

3. *Ziehl-Neelsen method.*

The staining fluid is composed by mixing:

Fuchsin	1 gram.
Absolute alcohol.	10 grams.
Five per cent aqueous solution of carbolic acid.	100 grams.

The different steps of the operation are exactly the same as in Lubimoff's method.

4. *Herman's method.*

The staining fluid is made extemporaneously and from two solutions:

(1) A one-per-cent aqueous solution of carbonate of ammonia;

(2) An alcoholic solution of methyl violet 6 B (1 of the violet to 30 of 95 per cent alcohol).

A few drops of the second solution are added to several cubic centimeters of the first until the mixture obtains a deep violet color and this is brought to a temperature approaching ebullition. The cover glasses or sections are left in the stain from one to two minutes, then placed for from two to three seconds in nitric acid diluted 1 to 4 for sections and 1 to 10 for cover glasses; after passing through the dehydrating fluids they are ready for mounting.

Double staining may be obtained by immersing the preparations, after passing through the nitric acid and the alcohol, in an aqueous or alcoholic solution of eosin.

5. *Kitt's method.*

This requires two fluids both of which preserve themselves indefinitely:*

(1) Aniline water,	100	grams.
1 per cent solution of caustic soda,	1	"
Fuchsin,	4 to 5	"
(2) Alcohol,	50	grams.
Water,	30	"
Nitric acid,	20	"
Methylene blue to saturation.		

* [The aniline-containing solutions will generally be found unfit for use after one or two weeks.—D.]

The preparations, cover glasses or sections, are left for from two to five minutes in the fuchsin solution heated as in the preceding methods, and then transferred to the second solution, in which they remain one to two minutes. During this last period all the elements other than the Koch bacilli are decolorized under the influence of the nitric acid, but, at the same time, fix the blue so that double staining is combined with decoloration.

The preparations are washed, dehydrated, and mounted.

All these methods have the same value in so far as the coloration which they give is concerned. In practice, however, those solutions may be recommended which are of longest preservation, and which require the fewest manipulations. The method of Kitt especially fulfills these two conditions.*

In the living subject about the only materials available for examination are fluid products, such as nasal discharge, milk or pus. In examining milk for tubercle bacilli the cover glass, after drying and before staining, should be freed from fatty matters by immersion in a mixture of absolute alcohol and ether, or in chloroform. If a deposit is formed at the bottom of the liquid this will furnish the best material for examination.

Very often the Koch bacillus has to be sought for in cadaveric products. In this case the investigations will be directed to the tubercular lesions or those sus-

*[Of the two methods most frequently employed—Ehrlich's and Ziehl's—the latter, on account of the greater preservation of the staining fluid, is often the most convenient, whilst the former gives a more brilliant color to the bacilli.—D.]

pected from their physical characters to be of a tubercular nature.

In the case of young, gray tubercles it suffices to spread upon a cover glass the product obtained by scraping their cut surface: the bacilli are here uniformly distributed. But when the tubercles are already caseated the caseous matter frequently contains but few bacilli except in birds, where, on the contrary, they are very numerous; this substance is first removed and the bacilli sought for in the wall of the caseous tubercle. The same method may be employed for cavities although these often contain a liquid very rich in bacilli.

Experimental inoculations.—Tuberculosis is inoculable to the horse, ass, ox, sheep, pig, dog, cat, rabbit, guinea pig, and to fowls.

Subcutaneous inoculation is without effect in the horse, ass, sheep, pig, dog, cat, and chicken.

Ingestion of virulent substances produces the disease, but not in all cases, in the horse, sheep, pig, dog, and cat. Cattle are easily tuberculized in this way. The chicken remains unaffected when sputa or tubercular products of mammals are mixed with its food.

Intravenous injection gives much surer results. Except in the case of birds it is almost invariably followed by a generalized tuberculosis of the lung with possible extension to other organs. In the ass the experiments of M. Chauveau have shown that these pulmonary granula heal spontaneously after a few weeks. Fowls do not usually contract the tuberculosis of mammals by way of the circulation. Thus, MM. Cadiot, Gilbert, and Roger have recently announced that out of forty pullets inoculated by them either in the veins or in the peritoneum, five only developed tubercular lesions.

The goat forms a specially unfavorable field for the development of tuberculosis and was long considered to be absolutely refractory to the inoculated as well as to the spontaneous disease. M. Nocard has recently described the evolution of the disease in a goat inoculated in the jugular five years before, and which at last had become affected with mange; M. Colin has also produced the disease in a goat inoculated under the skin with particles from the ox.

The laboratory animals—guinea pigs and rabbits—are very susceptible to the disease. The guinea pig is endowed with a quite special receptivity which makes it the reagent *par excellence* for tuberculosis.

Subcutaneous inoculation on the internal face of the thigh in the guinea pig is followed by a local abscess when the inoculated substance contains at the same time pyogenic germs, or only by a few yellowish granulations if it is pure; at the end of ten to fifteen days there supervenes an engorgement, sometimes an abscess, of the superficial inguinal lymphatic glands; the sublumbar glands of the corresponding side are invaded about the twentieth day; between the twenty-second and twenty-fifth days tubercles appear in the spleen and retro-hepatic glands; the lungs, liver, and the other lymphatic glands are attacked later. The disease lasts about two months. When the inoculation has been made at the ear the invasion takes place by the anterior lymphatics, and the lungs are attacked before the abdominal viscera.

Subcutaneous inoculation on the internal face of the thigh or at the ear, in the rabbit, does not give

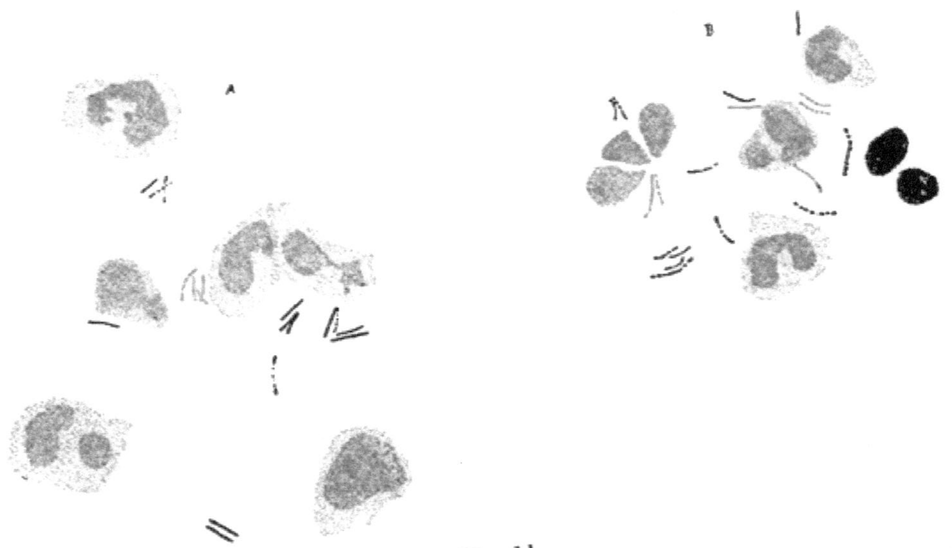

Fig. 11.

Tubercle bacilli; cover-glass preparations. Stained with Ehrlich's aniline-fuchsin. × 1000.
A. Pus of tubercular abscess in lung of cow. B. Human sputum in pulmonary phthisis.

rise to engorgement of the corresponding lymphatic glands. The local lesion is less pronounced than in the guinea pig, and generalization, less constant, occurs through the intermediation of the blood; the changes, here, are found chiefly in the lung. With bovine tuberculosis, in certain cases in the rabbit, M. Arloing has seen the development of glandular lesions resembling those of the guinea pig.

Intra-peritoneal inoculation in the rabbit and guinea pig determines tubercular lesions of the peritoneum, epiploic glands, liver, and spleen. The duration of the disease is always shorter than by the subcutaneous method.

Intra-vascular inoculation produces a generalized tuberculosis, but death is so rapid (fifteen to twenty days) that the specific lesions are not visible to the naked eye. The bacilli are disseminated throughout all the parenchymatous organs (septicæmic type, Yersin).

According to Straus and Gamaleïa this tuberculous septicæmia is obtained only with the cultures of avian tuberculosis.

Diagnosis of doubtful cases in cattle.—In cattle the diagnosis of tuberculosis by means of the clinical symptoms is often difficult. Not to speak of abdominal forms which are still more difficult of recognition, the discharge from the nostrils is often absent in pulmonary tuberculosis, and the search for the essential element—the bacillus—consequently denied to us.

M. Nocard, basing himself on the fact that cattle are accustomed to swallow their expectorations, advises looking for the bacillus in the pharyngeal mucus; this can be obtained by scraping the mucous

20

membrane of the throat with a spatula. Cagny proposes, in order to increase the bronchial secretion, the injection under the skin of ten to twenty centigrams of veratrine.

Poels has resorted, in the absence of discharge, to tracheotomy and the examination of the tracheal mucus.

The rarity of tubercular lesions of Demour and Decemet's membrane tends to refute the assertion of M. Mandereau as to the constant presence of the bacillus in the aqueous humor of tuberculous animals, and this assertion, which promised an easy diagnosis of the disease, has quickly been contradicted by Leclainche and Greffier, whose investigations on twenty animals which were certainly affected always gave negative results.

M. Peuch, having placed an irritating seton in a tuberculous cow, found, by inoculating the pus of the seton to guinea pigs, that the bacilli passed into the pus from the eighth to the fourteenth day. He therefore recommends the application of such an exudatory, and subsequent inoculation in order to remove the uncertainty of the diagnosis.

If the presence of the bacillus in the expectoration enables us to affirm the existence of the disease, its absence does not authorize us in positively affirming that it does not exist. Sometimes it is necessary to resort to inoculation of expectorated products, and the same precaution should be observed when dealing with milk, etc. Under such circumstances we have recourse preferably to the guinea pig, and only in default of this animal, to the rabbit.

When we possess pure products—cultures or young

tubercles collected in a state of purity and reduced to pulp—the inoculation can with advantage be made into the peritoneal cavity, thus obtaining a more rapid evolution; when we have only at our disposal virulent matters contaminated with other germs, such as pus or nasal discharge, we must content ourselves with subcutaneous inoculation. A region should be preferred in which the lymph glands are easily explorable in order that the progress of the lesions may be followed. In default of this exploration the emaciation of the subject after the time specified above will enable us to affirm, before the autopsy, that the inoculation has been successful.

Tuberculin.—Another means of diagnosis consists in testing with Koch's lymph or *tuberculin*.

Cultures of the tubercle bacillus contain a soluble product discovered by Koch which possesses a very remarkable property. This substance is without effect on healthy individuals, whilst it is toxic for the tuberculous. The name tuberculin has been given to a glycerin extract of cultures which contain this active agent. Since his first discoveries Koch has simplified the process by which it is prepared. Large cultures in veal bouillon, to which has been added one per cent of peptone and from four to five per cent of glycerin, are reduced to one-tenth their volume at a temperature of about 100°, then filtered through porcelain so as to remove all the microbes. The tuberculin thus obtained contains enough glycerin for its own preservation. The product has not a stable composition. Koch recommends testing its activity on tuberculous guinea pigs. According to this author, a good specimen of tuberculin, in the dose of one centigram, kills a guinea pig which has been inoculated eight to ten weeks before; it requires twenty to thirty centigrams, sometimes even as much as fifty centigrams, to kill a guinea pig inoculated four to five weeks before. The guinea pigs die in six to thirty hours, according to the extent of the tubercular process.

Tuberculin, after twenty-four hours' treatment with two to three volumes of alcohol, yields up its active substance in the form of a precipitate of an albuminoid nature.

Tuberculin causes in tuberculous subjects:

1st. A more or less intense febrile reaction which supervenes after several hours, usually from the tenth to the twentieth;

2d. A quite remarkable inflammatory reaction around the tuberculous foci. This substance is therefore pyretogenic and phlogogenic, but as the first of these actions produces its effects indirectly it is somewhat delayed. According to Gamaleïa, the mode of action of tuberculin is somewhat as follows: It possesses properties which are especially toxic for the elements of the tubercle, cause them to undergo necrobiosis in the same way as the secretions of the bacillus contained in the lesions produce necrobiosis (under the form of caseation or softening) of the central parts of the latter. Now, the proteins resulting from the decomposition of the elements thus attacked excite a local exudative inflammation and leucocytic infiltration. This local reaction results in the breaking up and elimination of the tubercular foci. The febrile reaction must be attributed to the absorption of the necrosed tissues. (See page 64, Gangolphe and Courmont.)

The special hyperthermic action of tuberculin on tuberculous animals renders it a diagnostic agent for tuberculosis.

The doses employed by different experimenters have varied within very large limits; in general, the injection of 20 to 40 centigrams of tuberculin is sufficient. From the tenth to the twentieth hour, occasionally sooner, an elevation of temperature of from one to three degrees is observed in tuberculous animals. This is therefore an excellent means of bringing to light obscure cases of tuberculosis. Unfortunately the rule is not without exceptions: some tuberculous subjects do not react at all and a certain number of others, not tuberculous, give the characteristic reaction. In spite of these exceptions tuberculin still furnishes us with a supplementary means of diagnosis and ought not to be discarded. M. Nocard has shown that it has no injurious action on lactation or gestation; he recommends its employment in the sanitary inspection of dairies where milk is produced which is intended for public consumption.*

On account of its phlogogenic and destructive action it might be supposed that tuberculin would act as a curative agent, but, unfortunately, this hypothe-

* [Tuberculin during the last three years has been very extensively tested in Europe and in America and the results obtained are still entirely in harmony with the opinion here expressed. Animals in the last stage of the disease do not react; those which from any cause have an abnormal temperature at the time of the inoculation are also unsuitable. When the test is carefully made the result is almost always reliable, although the extent of the thermal reaction gives no indication of the extent of the diseased process in the animal.—(D.)]

sis has not been confirmed; the numerous tests which have already been made have shown that tuberculin, instead of exerting a curative effect, may be positively harmful and cause the extension if not the generalization of the disease. The reagent is, in reality, without effect on the bacillus, whilst the inflammatory reaction which develops around the tubercles causes an aggregation of leucocytes which, becoming charged with microbes, transport these to points outside of the original lesion where they can incite new centers of disease; this inflammatory reaction can, further, become directly harmful when the tubercles are numerous and occupy a large extent of an important organ.

Etiology and pathogeny.—Contamination with tuberculosis is most frequently indirect, but may also take place in a direct manner.

Instances in which physicians and veterinarians have contracted the disease in making autopsies of diseased men or animals are incontestible, though fortunately rare. The virus inoculated through a wound, in such cases, occasions first of all a more or less limited cutaneous tuberculosis which may later become generalized.

Transmission of the disease from mother to fœtus constitutes another example of immediate contagion. This mode of transmission, although now placed beyond doubt both for animals and mankind, is actually of rare occurrence; the tubercle bacillus is in reality confined to the specific lesions and only exceptionally circulates in the blood; moreover, it has not yet been demonstrated that it is capable of passing through the villosities of the intact chorion, intra-uterine propagation appearing rather to depend upon some tubercular alteration of the maternal placenta. Johne has noted the existence of bacillar lesions in the liver and lung of a fœtus found in a phthisical cow. MM. Malvoz and Brouwier have communicated two cases of congenital tuberculosis in the calf; the first of these cases is quite convincing: the fœtus was removed from the healthy womb of a cow affected with the generalized disease; the second describes the case of a calf six weeks old, the origin of which was not determined, but in which the lesions were regarded by the authors as congenital because they were located in the same organs as in the first case, that is, in the liver, and hepatic and bronchial lymph nodes. From the absence of intestinal and pulmonary lesions it was inferred that the infection could only have taken place through the umbilical vein.

The presence of the bacillus having been demonstrated in the semen, some authors have been led to believe in the direct transmission from father to offspring by infection of the ovum. The special localization in the liver in well observed cases of congenital tuberculosis contradicts this manner of view.

The tubercular virus may be directly transmitted from a diseased to a healthy individual through sexual intercourse, either from the female to the male, or inversely.

Tuberculosis is usually communicated indirectly. The virulent matters rejected by the diseased (sputa of phthisical patients, nasal discharge and excrements of animals) and deposited on the ground, becoming dried and pulverulent, are carried with the air into the respiratory passages of healthy individuals, or are deposited on their food. Sputa or nasal discharge may also be directly ingested by animals; many instances of the propagation of human tuberculosis to the dog in this way have been recorded and it is probable that in the stables a considerable number of cattle contract the disease by consuming forage directly soiled by the expectorations of their neighbors.

The milk of phthisical cows appears also to be an important carrier of the germ; there is a difference of opinion on the question whether or not the udder can allow the passage of bacilli into the milk without itself becoming invaded by the tubercular process; some writers, basing themselves upon the result of numerous tests, have come to the conclusion that this does occur. However that may be, the difficulty of deciding as to the non-existence of tubercles in the mammary glands furnishes a sufficient reason for excluding from consumption all milk to which such suspicion is attached. Such milk is a dangerous food for human beings and also constitutes a source of infection for animals to which it is fed without previous cooking.

The flesh of cattle affected with tuberculosis and slaughtered for the market also becomes virulent under certain conditions but too little known. A certain number of inoculations which have been made with muscle juice coming from such animals have given positive results, although in the great majority of cases the results of these inoculations have been negative. Nevertheless the positive results obtained are sufficient to establish the possible danger of the flesh of phthisical cattle and to indicate the necessity of compulsory exclusion of such flesh from public consumption. This question has been lengthily discussed in the various congresses; without insisting further on the matter we will say that it is clearly connected with the question of the compulsory slaughter of all tuberculous animals. Indeed, before any diminution in the amount of flesh liable to seizure can be obtained the contagion must be checked between the living individuals. But, in the absence of any special provision of sanitary police in regard to such cases, the owners retain until the last stage, and in contact with healthy cows, those animals which, on account of their poor condition, they are unable to sell but the milk of which still secures for them a certain amount of profit.

We have now reviewed the different means by which the germs of tuberculosis are transported from diseased to healthy subjects. The receptivity of the subject plays an important rôle in the genesis of the process. It is very often dependent upon a special predisposition to the disease; this aptitude to contract tuberculosis may be acquired, in which case it results from the prolonged influence of bad hygienic conditions, or it may be transmitted to an individual by his ancestors. Heredity of the predisposition is a very common occurrence and one especially well recognized in the human family.

Acute or chronic catarrhal affections of the respiratory or alimentary passages favor the implantation of the tubercular virus either by diminishing the resistance of the tissues and of the organism or by producing solutions of continuity by which the germs may obtain entrance. The addition to pulverulent virus of bodies of an irregular shape, capable of wounding the respiratory mucous membrane, exerts a similar influence. Johne has demonstrated the great frequency of tuberculosis in cows which respire the fumes of iron foundries; these fumes are always loaded with fine metallic particles.

To the number of circumstances favoring the penetration of the tubercle bacillus into an organism must, further, be added the simultaneous presence of other microbes which prepare the field for the former. Associations of this kind may even start into renewed activity a tubercular focus which had long been latent and regarded as extinct.

The typical tubercular lesion, the tubercle, is not specific of the disease; it consists in an inflammatory nodule of peculiar structure which develops under the influence of multiple causes among which may be mentioned microbes (actinomyces, zoöglœa of Malassez and Vignal, pseudo-tubercle bacillus of Courmont, etc.), animal and vegetable parasites (demodex foliculorum, eggs of the strongylus vasorum, aspergillus, etc.), and non-vital agents (powders of cantharides, pepper, and lycopod, and croton oil). The tubercle bacillus must act by a mechanism similar to the preceding causes, that is, through the irritation which it excites in the elements which surround it. These elements are variable; they may consist of epithelial cells, endothelial cells, or the fixed cells of the connective tissue, and along with these should be mentioned the wandering cells or leucocytes which are every-where present and may play an important part. The cells in contact with the bacillus increase in size, their nuclei divide, and this is quickly followed by the division of the cell body. Simultaneously, the elements take a polyhedral form (epithelioid cells). This division continues and gives birth to a mass of new cells which constitutes the first stage of the tubercle. In the course of this initiatory period a certain number of elements may break up and disappear, but, as a rule, those which are directly in contact with the bacillar focus, around the center of irritation, acquire considerable dimensions, their nuclei continuing to multiply whilst the cell body remains single (giant cells); these giant cells, however, are also formed, at least under certain circumstances, by the fusion of several epithelioid cells. When the process begins in the interior of the blood vessels the leucocytes participate in it from the first, but in the opposite case they only intervene at a later stage by emigrating toward the invaded parts in order to enter into the struggle against the intruding germs; they then conduct themselves like the elements mentioned above. During this time there is formed at the periphery of the epithelioid zone a layer composed of leucocytes and fixed cells in way of multiplication, which in some degree limits or temporarily arrests the extension of the process.

The giant cells especially act as phagocytes; they contend against the bacilli, impair their vitality and tend to bring about their degeneration. Metschnikoff has described a series of involution forms of bacilli in the giant cells of the spermophile. The epithelioid and lymphoid cells have the same property. Both giant cells and epithelioid cells may be absent when the bacilli are extremely virulent or meet with an organism endowed with great receptivity; in such case these protective elements have not time to become developed and the tubercle is almost entirely lymphoid.

The extension of the inflammatory process to the capillaries involves their obliteration and the absence of the reparative plasma in the central parts of the tubercle. Thus is explained the necrobiosis (coagulative necrosis) so frequent in this lesion, although the influence of the secretions of the bacillus upon the organic elements must also be taken into account in this connection. This toxic action of the bacillus is the more pronounced in inverse proportion to the resisting power of the organism; thus, in spermophiles the giant cells of which destroy many bacilli and establish their superiority over them, caseous masses are not found even after a tuberculosis of long duration.

Tubercular lesions are usually localized at the point of entry of the germs; the disease is, therefore, primarily local, and it may definitely remain in this condition. More frequently, however, it extends, and this extension takes place in different ways. In man and the large animals extension occurs chiefly by the lymphatics, but it may also occur by way of the blood circulation when the bacilli have penetrated into the blood through ulceration of a vein at a diseased focus, or, indeed, through the medium of the lymphatic vessels.

We have already seen that the first mode of propagation predominates in the guinea pig, whilst it is only accessory in the rabbit.

Virulent products coming from one part of the organism may be transported to another part by the movements of the fluids in certain of the body cavities; virulent expectorations which are swallowed by the diseased animal can thus infect the digestive canal.

The localization of the lesions is dependent upon special predispositions of tissues or organs. This fact has been demonstrated by the experiments of Schuller, who showed that a contusion produced in one of the articulations in a subject artificially infected by any method whatsoever incites the evolution of a tubercular arthritis.

Tubercular lesions may become purulent in the absence of special pyogenic germs; the Koch bacillus, therefore, secretes a pyogenic substance. The injection of a culture in which the bacilli have been killed by heat is followed by an abscess.

Cultures sterilized and freed from bacilli are destitute of all power of producing tubercles in healthy individuals; when injected to tuberculous animals they give the reaction of Koch's tuberculin. The tubercular poison which occasions the specific neoplasms is, therefore, absent from the soluble portion of the cultures; according to Straus and Gamaleïa it exists in the bacilli themselves and persists after their death. Subcutaneous inoculation of dead bacilli causes a local abscess. Intra-peritoneal inoculation is followed by a tubercular peritonitis without other lesion. Venous injection results in a pulmonary tuberculosis in which the bacilli occur with their special staining characteristics unimpaired, but the tubercles thus obtained are not infective and have no tendency to become generalized. When the dose inoculated is sufficiently large they cause the death of the animal like the tubercles of living bacilli, and with the same general symptoms. Inoculation of a small dose of dead bacilli is followed by temporary loss of condition, and the subject becomes much more sensitive to a later inoculation with virulent bacilli.

Tuberculosis and scrofula.—The lesions of scrofula are of a tubercular nature; they contain the Koch bacillus.

M. Arloing has shown that the virus of tuberculosis and that of scrofula always infect the guinea pig, whilst the former only is virulent for the rabbit. He has also shown that local tubercular lesions are occasioned by bacilli which have become more or less attenuated; sometimes these local lesions are but slightly virulent and are of a scrofulous nature; sometimes they are more active, resembling tuberculosis properly so called, and like this have a greater tendency to extension.

M. Arloing is of opinion that the bacillus is attenuated in scrofula; M. Nocard, on the other hand, holds that the diminished activity of the scrofulous virus in the rabbit is due to the feeble receptivity of this species for tuberculosis and to the poverty of this virus in bacilli.

The virulence of scrofulous products is augmented by their passage through the organism of the guinea pig; whilst tuberculization is slow in the first guinea pig it becomes more and more rapid in the others inoculated later in the same series.

These observations concerning scrofula of man are

applicable to the scrofulo-tuberculosis of the pig (Nocard).

MM. Courmont and Dor have succeeded in producing local articular tuberculosis in the rabbit by intravascular inoculation of attenuated bacilli; one or several articulations were attacked, whilst the viscera remained unaffected.

Tuberculosis of mammals and avian tuberculosis.—The question as to the identity of these diseases has given rise to much discussion. We will describe first of all their differential characters:

The bacillus of avian tuberculosis is longer than that of the disease in mammals.

It grows on the different media when taken directly from the diseased animal, that of the mammalian disease only, in a satisfactory manner, after several transfers upon serum.

The vegetation of the bacilli of fowls is more rapid; their cultures on solid media are thick, moist and luxuriant; those of the bacillus of man are meager, dry, scaly and of dull appearance.

Cultures derived from fowls preserve their vitality longer (ten months at least) than those coming from man (six months).

The avian bacillus grows at a temperature as high as 43°; that of man ceases to grow at 41°.

Avian tuberculosis is with difficulty transmitted to the guinea pig and the lesions do not become generalized as in that which is derived from the human being.

The rabbit, although no more sensitive to the tuberculosis of fowls than to that of mammals, shows a much great susceptibility to the former than the guinea pig. According to Straus and Gamaleïa, avian tuberculosis gives rise only to Yersin's septicæmic type of the disease when it is inoculated in the veins of the rabbit.

The dog readily contracts human tuberculosis; it does not take the tuberculosis of fowls.

Fowls are refractory to human tuberculosis whilst very sensitive to inoculations of avian tuberculosis. The lesions become localized in the abdominal organs, occasionally in the lungs, often also in the marrow of bones (causing lameness).

The question is whether or not these differences are sufficiently important to justify us in regarding the two bacilli as distinct species.

Both bacilli have the same form; the differences in size which have been observed are of no importance; long bacilli may be seen in mammals and short bacilli in fowls. Moreover, both are liable to vary in their cultures and according to the animals to which they are inoculated; we have seen the bacilli of avian tuberculosis become exceptionally short when inoculated to the calf.

Both bacilli behave alike toward coloring matters. They produce lesions showing the same structure and the same general evolution.

The differential characters drawn from cultures are not absolute.

The inoculation of avian tuberculosis to the guinea pig occasionally causes a generalization quite similar to that regularly occasioned by human tuberculosis. Out of twenty-seven guinea pigs inoculated with the spontaneous lesions of the former disease five showed a local abscess, seven a discrete visceral tuberculosis, and two the generalized disease (Cadiot, Gilbert and Roger). Avian tuberculosis which had acquired greater virulence by passing through the organism of the rabbit killed seven out of eight guinea pigs which received it in the subcutaneous cellular tissue (Courmont and Dor). In this case, therefore, the bacillus from fowls behaved exactly like that from mammals.

According to an experiment of our own the avian bacillus inoculated in the cellular tissue of the calf causes a disease exactly like the human bacillus; as with the latter, the evolution is slow and remains for a long time limited to the glands receiving the lymph from the place of inoculation.

The non-receptivity of fowls for mammalian tuberculosis is not absolute; out of numerous tests which have been made a certain number have given positive results; the lesions, however, were always less generalized than with the avian tuberculosis.

Bacilli derived from avian tuberculosis, and which had not passed through fowls for at least five years, were found to have become more active for mammals and produced a generalized tuberculosis in rabbits, guinea pigs and chickens. From the liver of one of those chickens four guinea pigs were inoculated, none of which became tuberculous. A single passage through the chicken, therefore, was sufficient to impair the virulence for mammals of these avian bacilli which previously had acquired a virulence approaching that of the human specimen.

The bacillus of fowls becomes more active for mammals and less active for fowls by passing through the organism of mammals. This fact has been proved by

MM. Cadiot, Gilbert, and Roger by means of the generalized lesions which they obtained in two guinea pigs with virus coming from the pheasant. After three passages through mammals this virus was inoculated without effect to two chickens.

The injection into animals of tuberculin prepared from the bacillus of fowls is followed by the same effects as the injection of tuberculin of human origin.

We must therefore regard the bacilli of avian tuberculosis and the bacilli of mammalian tuberculosis as varieties of the same species.

Vaccination.—Tuberculosis is essentially a recurrent disease; a first attack begets a predisposition to a second. It seems, therefore, *a priori*, paradoxical to endeavor to prevent it by means of culture products. Nevertheless, MM. Richet and Hericourt, Courmont and Dor, have succeeded in producing in the rabbit a certain degree of immunity against tuberculosis by the injection of sterilized cultures of the bacillus of avian tuberculosis; they employed bouillon cultures and obtained sterilization by heat or by filtration. The disease was retarded in the majority of the vaccinated animals; in some it was completely prevented. These results justify us in affirming the existence of vaccinating substances in cultures of the tubercle bacilli and of indulging the hope that, some day, it may be possible to vaccinate the human being against this terrible disease.

Vaccination experiments have been made in the rabbit by means of the blood serum of the dog (hæmocyne). The investigators who have taken the initiative in these experiments seem to have obtained if not absolute immunity at least a retardation of the evolution of the experimental disease. The blood of dogs previously tuberculized possesses properties more active in this respect than that of healthy dogs. These results have been utilized in therapeutics, hæmocyne having been employed in the treatment of human tuberculosis. Tests of the same kind have been made with the blood of the goat.

Microbic tuberculosis other than that of Koch.—We have already said that microbes other than the Koch bacillus are able to engender the special inflammatory reaction which is characteristic of tubercle. Of these we are acquainted with several; we will only refer specially to the tubercle-begetting zooglœa of Malassez and Vignal, and to the bacillus of Courmont.

Zooglœic tuberculosis of Malassez and Vignal.—In studying experimental tubercular lesions of the guinea pig the authors found, in place of Koch's bacillus, micrococci associated in zooglœa. These germs are not stained by Ehrlich's method; they may be demonstrated by the following process:

Sections are left during two or three days in a mixture of:

Two per cent solution of carbonate of soda, ...	10 volumes.
Saturated aniline water, . . .	5 "
Absolute alcohol, ...	3 "
Solution made with 9 vols. of distilled water and 1 vol. of concentrated solution of methylene blue in 90 per cent alcohol, ...	3 "

They are washed with water, dehydrated in absolute alcohol colored with methylene blue, and cleared in oil of bergamot or turpentine.

This solution is a combination of the two solutions, staining and decolorizing, of Malassez and Vignal's original process (see page 110).

The bacterial elements are short, rounded or slightly elongated, associated in chains or in small groups, or more frequently in large groups or zooglœa.

Zooglœic tuberculosis is transmissible from guinea pig to guinea pig by inoculation and gives rise to generalized lesions like the tuberculosis of Koch; but death supervenes at the end of six to ten days, that is, much more rapidly than in the case of the latter.

The authors were inclined to think from their first observations that the zooglœa and the Koch bacillus were the same micro-organism under two different forms; now, however, they must be regarded as absolutely distinct germs.

M. Nocard had the opportunity of studying an enzoötic of zooglœic tuberculosis in which all the chickens of a farm succumbed to the disease. The tubercles, in all cases, had their seat in the lungs, a situation in which they are hardly ever seen in genuine tuberculosis. Moreover, Koch's bacillus was absent, whilst the zooglœæ were abundantly represented in the lesions.

MM. Nocard and Masselin produced zooglœic tuberculosis by inoculating the guinea pig with the nasal discharge coming from a cow suspected of phthisis, and which, at the autopsy, was found to be exempt from this disease. The cocci found in the guinea pig were cultivated and successfully inoculated to the guinea pig and the rabbit.

Bacillar tuberculosis of Courmont.—M. Courmont found in tubercular lesions of the pleura in an ox—lesions which did not contain Koch's bacillus—a short bacillus with its substance condensed at the two extremities and with a clear slightly constricted median zone; this bacillus is never associated in chains or in diplo-bacilli; it is aërobic and anaërobic. It is easily cultivated and grows rapidly in all the culture media and at wide limits of temperature, even up to 46°.

Guinea pigs succumb in four to eight days with a local œdema and great enlargement of the spleen, but without tubercular lesions. The bacilli are abundant in the serosity of the œdema and in the blood; after several passages through the guinea pig a caseous abscess develops at the place of inoculation.

Rabbits contract a more or less complete tuberculosis; an abscess forms at the place of inoculation, and, after death, disseminated or confluent tubercles are found in the spleen, liver, and lungs. These tubercles have the classic structure; they do not contain Koch's bacillus but those described above.

A culture twenty days old having been inoculated to guinea pigs the latter died in less than ten days with a generalized tuberculosis in which the lesions contained Courmont's bacillus. The property of begetting tubercles in the guinea pig appears to exist in cultures only from the twentieth to the twenty-fifth day; at other times the inoculated guinea pig dies without tubercles. The tubercles of the guinea pig kill the rabbit, but without tubercular lesions.

Courmont's bacillus, therefore, produces a tuberculosis in the ox; the lesions of the ox have direct tubercle begetting action on the rabbit but not on the guinea pig, except under the special conditions mentioned above. We know, however, that the guinea pig is more sensitive to mammalian tuberculosis than the rabbit, hence, besides the character of the bacillus, there is a manifest difference between the two diseases. Furthermore, the experimental tuberculosis obtained in the guinea pig by the subcutaneous inoculation of Koch's bacillus develops so slowly that, on an average, the lung is invaded only at the end of two months; with the bacillus of Courmont, as also with the zoogleæ of Malassez and Vignal, the guinea pig becomes tuberculized in less than ten days. Generally, in the rabbit, the duration of Courmont's tuberculosis is almost the same as that of Koch's tuberculosis. Another character important to notice is the presence of Courmont's bacillus in the blood of subjects tuberculized by means of this bacillus. Finally, the disease generalizes without infecting the lymphatic glands.

Glanders.

Glanders is an infectious and contagious disease, with progressive course, characterized by circumscribed and multiple alterations localized in the respiratory mucosa (glanders), or in the skin (farcy). These two localizations may coexist, in which case we have to do with the disease glanders-farcy. The disease is peculiar to solipeds: horse, ass, and mule; it may also develop in man as a result of accidental inoculation with virulent products coming from one of these animals; transmission to the lion, dog, and goat has also been observed as a result of spontaneous contamination.

The disease is acute or chronic; it is always acute in the ass and mule, as well as in the lion.*

Microbe.—The bacillus mallei is a slender, motile rod, straight or slightly curved, and with rounded ends. It measures 2μ to 5μ in length by $0{\cdot}5\mu$ to $1{\cdot}5\mu$ in thickness, hence it is thicker than the Koch bacillus, but of the same length. When stained it shows alternating clear and colored spaces which give it a granular aspect recalling that of the tubercle bacillus. Rosenthal regards these clear spaces as spores; others, however, basing themselves on the slight resistance of the bacillus to heat, refuse to admit the existence of spores.(1)

Fig. 12.

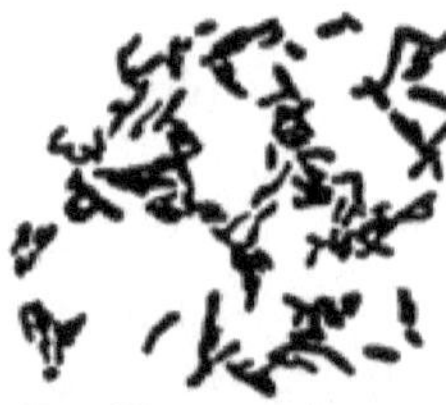

Bacillus mallei. × 1,000. From a photomicrograph. (Fränkel and Pfeifer.) — From Sternberg's Bacteriology.

The glanders bacillus is aërobic; it occurs in the pathological secretions—nasal discharge, pus of ulcers; and in the specific lesions—farcy buds, tubercles, glanderous ulcers, and inflammation of the corresponding lymphatic vessels and glands, etc. It is

(1) Babès has shown that such granules are present in the majority of bacilli, even the least resistant, and that they do not possess the resisting power of true spores. He believes that they exercise the same function in the multiplication of bacilli as the chromatic part of cell-nuclei in the division of cells.

* [The sub-acute or chronic form of the disease seems to be not uncommon in mules in the Southern States.—D.]

+ present in the blood only in acute forms, and then in very small numbers.

Action of physical and chemical agents.—The contagion of glanders is destroyed by two minutes exposure to a temperature of 100°, but its destruction can be obtained at lower temperature, provided the action of the heat is more prolonged, for example, five minutes at 65°, ten minutes at 55°. Glanderous pus, spread in a thin layer and left to desiccate in contact with the air, loses its activity between the second and third days; hot and dry weather favors its destruction, while cold and wet weather retards it. Under the same conditions of desiccation, but excluded from the air, it yet shows itself active after twenty-six days. Virus rapidly and thoroughly dried retains its vitality in contact with the air longer than that which is slowly and imperfectly dried. The discharge from the nose in glanders, when immersed in water, has been found to retain its virulence for eighteen days. Virulence is not readily destroyed by putrefaction; inoculations made with the central part of pieces of glanderous lungs abandoned to the air for fifteen, eighteen, and even twenty-six days, have given positive results. (Cadeac and Malet.)

The following substances destroy the virus of glanders after one hour of contact: carbolic acid, 2 per cent; sulphuric acid, 2 per cent; chloride of zinc, 2 per cent; saturated lime water; hypochlorite of lime, 1 per cent; corrosive sublimate, 1 to 1,000, and 1 to 10,000; sulfate of copper, 5 per cent; permanganate of potash, 5 per cent; nitrate of silver, 1 to 1,000; chlorine gas, and sulphurous acid gas, in concentrated solution.

Cultures.—Cultures are easily obtained on the different media; they require free access of atmospheric oxygen; 37° is the most favorable temperature; below 20° their growth is arrested except upon glycerin-agar; it also ceases at 43°, and the germs are killed at 55°.

Bouillons become turbid within twenty four hours without presenting special characters.

On agar and on serum it shows a bluish-white translucid growth in the form of droplets, or in a continuous layer which becomes opaque as it increases in thickness.

Potato is admirably adapted to the culture of the bacillus of glanders. On this medium it forms a thick, moist, glistening, viscid coating, which after a few days assumes a fawn color, gradually deepening to a bright chocolate. (Nocard.) This culture is characteristic and should assist us in the diagnosis of doubtful cases; it suffices to sow a particle of the suspected product, previously diluted, upon potato; the latter soon becomes covered with diverse growths among which we should recognize by their peculiar color those due to the glanders bacillus.*

Research and coloration.—The *bacillus mallei* shows little affinity for the aniline colors; it does not support the Gram or Weigert stains. The methods

* [Such simple culture tests can be of value when the suspected material is obtained from a yet unopened skin nodule or from an extirpated submaxillary lymph node, but will rarely assist us in the examination of nasal discharge.—D.]

of double staining are therefore inapplicable. Generally we have recourse to Löffler's blue, decolorizing with carbolated water, at 1 to 300.

Kühne also recommends the following method of staining: Sections, well freed from alcohol, or cover glasses, are placed in a solution composed of: water, 100; carbolic acid, 5; alcohol, 10; methylene blue, 1·5. They are decolorized by a rapid passage through water acidulated with hydrochloric acid, and washed in distilled water. The sections are dehydrated by a short immersion in alcohol, then placed in aniline oil to which has been added a few drops of oil of turpentine, then in pure turpentine, and, finally, in xylol.

The bacillus of glanders is difficult to demonstrate in old lesions, where it appears to break up into granules (perhaps into spores); it is always much more abundant in acute lesions.

Experimental inoculations.—Solipeds, the sheep, goat, pig, dog, cat, rabbit, guinea pig, field mouse and pigeon take the disease by inoculation. The ox, white mouse, rat and chicken are refractory. The ass always contracts acute glanders by inoculation. When the virus is inserted by scarifications in the forehead, for example, an extensive ulcer, with indurated and inflamed borders, develops on this region. Along with this the manifestations of the general disease appear about the third day, and the subject quickly dies. The autopsy reveals an eruption of small reddish nodules on the respiratory mucous membrane, pyæmic infarcts in the lungs, liver, kidneys, spleen, marrow of bones, etc.

In the dog the insertion of the virus in the skin or

the subcutaneous cellular tissue is sometimes followed by a violent access of fever in the course of the third day; then the local lesions appear in the form of an extensive engorgement which bursts and ulcerates, the chancre thus produced having a great tendency to spread. Farcy buds and cords, and rodent ulcers often appear on other parts of the body; sometimes an intense and painful lameness—expression of glanderous arthritis—supervenes without apparent cause, the stifle being the most frequent seat of the specific inflammations. Out of twelve dogs inoculated by Prof. Reul, four died, three were killed, and five spontaneously recovered. Inoculation of glanders to the dog often gives, as sole reaction, a superficial ulceration which cicatrizes in eight to fifteen days.

In the guinea pig inoculation by scarifications (these are made on the neck or back) is followed by ulceration of the wounds on the fifth to the tenth day; these ulcers may, later, cicatrize. Subcutaneous inoculation gives voluminous abscesses in the whole chain of lymphatic glands proximal to the place of inoculation. In both cases the animal becomes emaciated and dies more or less rapidly with tubercular foci in the spleen, liver, lung and lymphatic glands. The local lesion—the ulcer—suffices to establish the glanderous nature of the inoculated products. When the inoculation is made in the peritoneum a pronounced swelling of the scrotum is observed from the second to the third day, this swelling indicating the specific inflammation of the testicular membranes. This peculiarity can be utilized for the purpose of making a rapid diagnosis of the disease (Straus).

When glanders is inoculated to the dog or guinea pig by scarifications it may be followed by a negative result. Hence, an absolute value should be assigned only to positive results, and the suspected material which has been unsuccessfully inoculated submitted to further tests.

Diagnosis of doubtful cases.—Besides experimental inoculations to susceptible animals, guinea pigs, dog, and ass, and the cultures on potato of the suspected discharge, some have counseled the employment of auto-inoculation. This operation consists in inoculating a horse with the products, supposed to be virulent, coming from the same animal; these insertions are made in the skin by puncture or superficial incision and the existence of glanders confirmed if the operation is followed by the formation of a chancre; but, as we shall see later, no conclusion can be reached founded on the absence of the ulcerous reaction. In default of any discharge some have resorted to extraction of a swollen lymph node and examination of the same for the specific bacilli. Finally, we may have recourse to test inoculations with mallein.

Mallein.—By this name is designated a glycerin extract of pure cultures of the bacillus of glanders. This is the glanders lymph, the analogue of the tuberculosis lymph or tuberculin. The extract is sterilized by heat, and diluted to ten times its weight with two per cent carbolic solution.

The injection of 30 to 50 centigrams of this dilution to glanderous horses produces a reaction characterized especially by dejection, acceleration of the pulse, and elevation of temperature ranging from 1·5° to 2°, by a hot œdematous swelling as large as the hand at the place of inoculation, and by an appreciable swelling of the subglossal lymphatic glands. In a healthy animal the same injection would produce no effect. Mallein should, therefore, facilitate the diagnosis of the disease in doubtful cases.

Etiology and pathogeny.—The efficient cause of glanders is the *bacillus mallei.* This microbe, under natural conditions, multiplies only in the organism of solipeds and its origin must be looked for in these animals. The virulent substances cast off by a glanderous subject are : nasal discharge, pus of ulcers, saliva, urine, pus of setons, and semen ; the virulence of the last four products, though less constant than that of the first, is nevertheless incontestable and has been demonstrated by experiment. To this list must be added, in the case of the cadaver, the various specific lesions and the muscles. Inoculation of guinea pigs with muscle juice has produced the disease in a certain number of cases.

Contagion takes place by direct or indirect contact. There is direct contact from the horse to man when the latter inoculates himself in manipulating or in dressing glanderous lesions; from horse to horse when two horses, one of which is glandered, occupy adjacent stalls in a stable, or work side by side so that they can easily touch each other. The disease is occasionally transmitted through sexual intercourse ; Zundel mentions the case of a glandered stallion which infected more than fifty mares. The eventual passage of the glanders bacillus into the semen accounts for this mode of infection, whilst their filtration through the placenta occasions transmission of the disease from mother to fœtus, an occurrence which has been observed by MM. Cadeac and Malet. Inter-sexual and intra-uterine contamination are, however, actually of rare occurrence.

Indirect contagion is much the most frequent. The glanders virus, distributed externally, contaminates the food, drinking water and litter, harness, grooming utensils, sponges, brushes, curry-combs, etc., the walls of houses, mangers, racks, and such like; if it meets with the conditions necessary for desiccation it will be carried with the dust into the surrounding atmosphere. The morbific germ is transported on to the healthy organism through the intermediation of numerous vehicles. Infection may take place through the respiratory passages, the animal inspiring air charged with virulent particles, but this mode of contagion should be rarely effective in glanders since desiccation is a puissant cause of the destruction of the bacillus. The disease is more certainly communicated by the digestive canal, through swallowing infected food or water. This contamination of the ingesta may result from their immediate contact with the products coming from a diseased animal or it may occur during their storage in infected places.

In this connection should be mentioned the danger arising from the consumption of animals which have died or been killed while suffering from the disease. This danger exists not only for the specific lesions, in which the virulence is evident, but also for the flesh. The investigations of MM. Cadeaç and Malet have shown that the juice of such flesh is capable of communicating glanders.

Finally, the virus may penetrate through the skin

to which it has been carried in various ways and especially by the harness, grooming utensils, litter, etc.; it gains entrance through accidental abrasions of the integument. According to Babès, however, the glanders virus can make its way into the organism through the intact skin, penetrating the orifices of the hair follicles. This penetration will by facilitated by frictions. M. Nocard undertook to test this assertion by rubbing an ointment charged with glanders bacilli on the skin of three asses and fifteen guinea pigs. Of these, only two guinea pigs became glandered, a result which considerably reduces the risk which might be inferred from the above conclusions.

Certain circumstances are of such a nature as to favor the implantation of the bacilli of glanders. We recognize the predisposing influence of bad hygienic conditions, excessive fatigue, and chronic exhausting diseases. Chronic forms of glanders may assume acute characters under the same influences.

The bacilli of glanders having once obtained entrance into the economy multiply generally at the place of their penetration, thus producing local lesions. They quickly invade the lypmphatic system (which becomes the seat of specific inflammations, glanderous lymphangitis and adenitis) and the blood; by the latter they are carried throughout the system but they only develop in the tissues predisposed to their attack: respiratory mucous membrane, integument, testicles, and synovial membrane of articulations and tendons, etc.

The specific lesions present certain analogies with those of tuberculosis. The primary lesion—the glanders tubercle—is purulent at its center and destitute of giant cells; besides this marked tendency to suppuration, should be mentioned the early retrogressions which lead to the ulcerations characteristic of the disease.

Vaccinations.—Glanders leaves behind it no immunity against another attack. Thus, auto-inoculations and re-inoculations performed upon animals already affected with the disease are followed by positive results. In the glanderous guinea pig these inoculations determine local symptoms as intense and a generalization as complete as at the time of a first insertion; in the dog the second attack is also generalized and often as severe as the first. According to Galtier the dog can contract the disease as often as five times, but the local symptoms become less and less marked In the horse re-inoculations and auto-inoculations may prove abortive or may produce a chancre with or without lesions of the lymphatics, but fever and aggravation of symptoms are never observed, the virus seeming to limit its effects to the point of inoculation. In the ass secondary inoculations are followed by much more intense reactions; each insertion produces a marked tumefaction and a corresponding ulcer, corded lymphatics, glandular engorgements—in brief—the effects of a first inoculation. (Cadeac and Malet.)

The natural disease, therefore, does not confer immunity. Several attempts at artificial vaccination have been made. Straus obtained immunity in a dog by the intravenous injection of a small quantity of virulent culture. A benign disease is produced which immunises the dog against intravenous inoculation of large doses.

Epizoötic lymphangitis, or African farcy.

Rivolta has described, and M. Nocard confirmed, the presence in the pus and lesions of African farcy of "a sort of micrococcus, slightly ovoid and somewhat pointed at one of its extremities, measuring 3μ to 4μ in diameter; its contour is clearly defined by a very refringent line." This organism (cryptococcus of Rivolta) is colored by the Gram-Weigert-Kühne methods; but its dimensions and its refringence are such that it is impossible, according to M. Nocard, to confound it, even when unstained, with any other element.

Several practitioners have described the appearance of chancres of acute glanders on the nasal mucosa of animals attacked by epizoötic lymphangitis. The demonstration of the cryptococcus in these lesions enabled M. Nocard to affirm that they were related to lymphangitis and not to glanders. Moreover, the bacillus of this last disease was lacking.

*Strangles.**

We are indebted to Schütz for the investigation of the microbe of strangles. It is a streptococcus which occurs in short chains, in diplo- and in monococci in the nasal discharge and in the pus of the lymph-glandular inflammations and abscesses symptomatic of the disease; in long, tortuous chaplets in sections of the inflamed organs. It readily takes the different aniline stains. Inoculated under the skin of the

* [Also called, in America, "Distemper" of horses; *Fr.*. Gourme; *Ger.* Druse.—D.]

horse it causes the formation of an abscess. In the mouse* it produces, in addition, metastases by way of the lymphatics and blood vessels. Natural infection takes place through different channels but more especially through the respiratory and digestive mucous membranes. Its absorption is facilitated by the presence of solutions of continuity, but it has not been shown that these are necessary. Accidental or operative wounds also sometimes furnish ports of entry for the virus; castration wounds, for example, may be infected by the surgeon himself, his hands or instruments being soiled with the germs from previously handling horses suffering from strangles. The disease is transmitted from the mother to the fœtus. Whatever may be its mode of

Fig. 13.

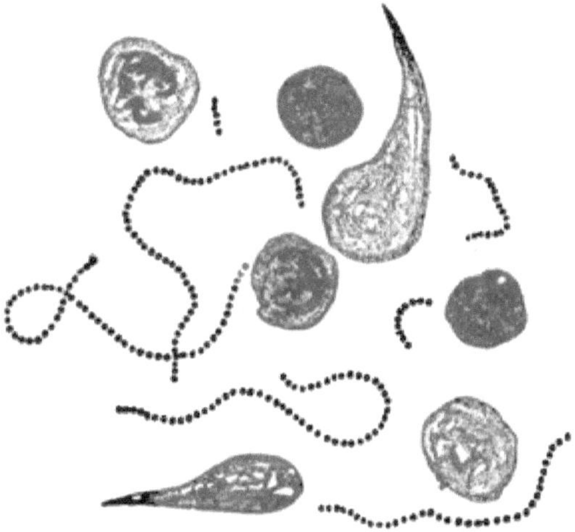

Streptococcus equi, pus of strangles of horse. × 1000. (Kitt.)

* [An important differential test in distinguishing this disease (in its less characteristic forms) from glanders consists in the inoculation of mice. White and gray house mice are highly susceptible and die from inoculation with strangles, but are insusceptible to glanders. Field mice, on the contrary, die from inoculation with glanders, but develop only a local lesion from inoculation with strangles.—Th. Kitt: *Bacterienkunde, p.* 254.—D.]

entry the germ appears to pass at once into the blood as is indicated by the high fever at the beginning of the disease; it then becomes localized in the lymphatic system. The microbic theory of strangles accounts for its contagiousness, the secondary abscesses met with in the course of the disease, and the septico-pyæmic complications which, in certain cases, terminate it.

Contagious acne of the horse.

This disease is characterized by a pustulous dermatitis which is very readily transmitted to other animals. It is generally benign but may be more severe and lead to the production of ulcerations and inflammations of the lymphatic vessels and glands. It is, however, always easily distinguished from farcy by the great tendency to cicatrization of the ulcers which it occasions; in farcy, moreover, true pustules do not occur. Acne is distinguished from horse-pox or variola of the horse by the fact that the eruption in this last disease is always localized in the lips, nostrils, and pasterns.

Dieckerhoff and Grawitz discovered in the pus of acne a short bacillus (2μ) which they cultivated and the culture of which, rubbed into the skin of the horse, reproduced the disease. The guinea pig succumbs in twenty-four hours when subjected to the same treatment. These cultures are also pathogenic for the ox, sheep, dog, and rabbit. Inoculated under the skin of the mouse it gives rise to the formation of an abscess.

The natural disease is transmitted by means of the grooming utensils, harness, blankets, etc.; thus, it is not uncommon to see it localized on the regions covered by the saddle or girth.

Actinomycosis.

Actinomycosis is a disease, most frequently of a local character, caused by a vegetable parasite, the actinomyces. The lesions by which it is characterized have long been recognized in practice, but their true nature was entirely unknown before the investigations of Bollinger, the different names, such as *osteosarcoma*, *osteoporosis*, *spina-ventosa*, *cancer or farcy of the bones*, etc., by which the maxillary tumors of cattle were designated reflecting the very diverse and somewhat vague opinions of the various writers upon these productions.

The discovery of a special fungus in the majority of these tumors has enabled us to classify them with the parasitic infections and explain their great intractability.

The facts established by the authors in their first communication on actinomycosis have been repeatedly confirmed since then in both human and veterinary medicine, and, referring here to the latter branch of medicine only, the lesions of this disease have been found in other parts of the body than in the maxillæ. They have been met with in the tongue, pharynx, reticulum, liver, nasal cavities, larynx, lungs, neck, and vertebræ.

In the pig they have been observed in the muscles, lungs, amygdalæ and mammæ; in the horse in certain cases of scirrhous cord. A case has also been recorded in the dog.

Characters of the parasite.—The actinomyces (ray

fungus) occurs in the specific tumors in the form of yellow or occasionally colorless grains; when the material in which they are contained is spread out on a glass slip the smallest of these grains appears somewhat like grains of sand; the larger are formed by the union of primary granules and are of various forms. These grains are most frequently cretaceous, having a hard, stony consistence. In size they vary from 0·1 millimeter to 1 millimeter, or more.

They are composed of one or several colonies of a fungus the elements of which are arranged in rays. In each colony there may be distinguished:

1st. A central zone, formed of very fine filaments ramifying and intermingled in a close felted net-work. The diameter of these filaments is uniform in all parts of the central zone; in structure they consist of hollow cylinders, each, at intervals, containing a nucleus which readily takes up coloring matters.(1)

The size of this central part of the actinomyces corresponds with that of the tuft or colony; small in microscopic grains it is of much larger dimensions in those tufts which have acquired a considerable volume.

2d. A peripheral zone, rendered conspicuous by the radiating elements of which it is composed; these elements are pyriform with their large extremities losing themselves in the tissue which surrounds the colony and even occasionally penetrating into the adjacent cells, and their slender ends passing into the

(1) Actinomycosis and its parasite. See *Annales de médicine vétérinaire*, 1890.

central zone with the filaments of which they become continuous.

The thickness of the peripheral zone is quite uniform whatever may be the size of the actinomyces itself, but it may be unequal in the different parts of the same colony on account of the varying dimensions of the cortical enlargements; the average length of these enlargements is from 15μ to 30μ and breadth 5μ to 7μ, but these limits may be much surpassed; we have seen some which measured 74μ in length by 10μ in thickness.

The club-shaped enlargements may be simple or branching, and branching may take place either from the slender pedicle or from the enlarged part itself; the branches, emerging in this manner, may themselves divide giving rise to new club-shaped swellings, so that the whole obtains a more or less dense arborescent appearance. These club-shaped enlargements are composed of a resistant membrane and clear contents; occasionally the membrane shows circular depressions which seem to divide the body into small cubical elements, and at these points transverse division is readily produced.

Besides the typical actinomyces just described there are others which are quite small and in which the enlargements are absent, and others, again, in which the filamentous central zone seems to be entirely transformed into these club-shaped elements.

In colonies of large size the central zone contains micrococci, appearing like small round points of less than 1μ in diameter, united in chains, or, more frequently, in small irregular masses; these nests of micrococci, in the largest tufts, are not infrequently found in large numbers, their nature being more easily recognized by the rarefaction of the mycelial felt-work in these situations.

Fig. 14.

Tuft of Actinomyces. (From Kitt's *Bacterienkunde.*)

Isolated clubs of Actinomyces. (Johne.) Do.

MM. Cornil and Babès have described a special condition of the filaments of the periphery of the colonies, in which they terminated by various slight enlargements bearing conidia.

Harz classes the actinomyces with the hyphomycetes fungi, regarding it as a complete fungus composed of mycelia, hyphæ and spores, the *mycelium* being represented by a basal cell from which spring the *hyphæ*—the branching filaments of the internal zone—and these bearing the spores or enlargements at the periphery.

The basal cell of Harz has not been found by other investigators; most of these, however, agree in regarding the enlargements as spores or, rather, as sporangia; hence these enlargements have received the name of *conidia.*

The classification of the actinomyces with the fungi appears to us to be supported only by the results of the examination of this parasite in the tumors to which it gives rise.

Bostroem classes it with the group of schizomycetes and especially with the cladothrix, a conclusion which is forced upon us from a study of the parasite in its cultures, and one to which our own researches have led us. We are also inclined to regard the micrococci, spoken of above, as spores.

The actinomyces is anaërobic and facultatively aërobic.

Action of physical and chemical agents.—According to Domec, the filaments are killed in five minutes by moist heat at 60°, whilst the spores resist for that length of time a temperature above 60° but below 75°. This feeble resistance of the spores compared with that of the spores of bacteria has been advanced as an argument in favor of regarding these spores as allied to those of the mucedineæ, and of classifying the parasite with this group of plants.

The action of chemical agents has been little studied. Iodine has been recommended for the destruction of the actinomyces in the tumors, but the hot iron is still the means most to be recommended for this purpose when it is possible to use it.

Research and coloration.—The examination of the actinomyces fungus is extremely simple, the suspected material merely requiring to be spread upon a glass slide; the hard, yellow grains which to the naked eye have the appearance of grains of sand, separate themselves during the operation. These grains are the colonies of the actinomyces. To verify this it is merely necessary to put on a cover glass and examine with the microscope; a magnification of from two to three hundred diameters will always prove sufficient for the recognition of the parasites.

This mode of preparaton, however, is inadequate for the detailed study of the fungus. The center of the radiating masses is nearly always calcified, and this transformation conceals its filamentary structure and, at the same time, prevents the dissociation of the tuft. The latter is also enveloped by a mucoid, viscid substance which tends to hold together the enlargements at the periphery. The inconvenience arising from these two circumstances is obviated by treating the substance to be examined with alkalies or dilute acids.

For this purpose we have found the employment of dilute ammonia quite satisfactory; the sand-like grains yet require to be crushed in order to spread, or better, to dissociate them.

By this means beautiful preparations are obtained in which the mycelium and its connection with the conidia can be studied.

The examination of sections is highly instructive. The actinomyces can be stained in different ways. According to the way which is mostly recommended, the sections are immersed in a solution of orseille,(1) then in alcohol, and, finally, in an aqueous solution of gentian violet or methylene blue. A double staining is thus obtained, the conidia being red and the mycelium violet or blue.

In our opinion Weigert's method gives better results. The antinomyces take a violet color and the

(1) Pure orseille, free from ammonia, dissolved until a deep red color is produced, in: acetic acid, 5; absolute alcohol, 20; water, 40.

tissues of the tumor may be stained red by picro-carmine.

The methyl violet, by this method, may act in two degrees. When the preparation has been exposed to the staining solution for a short time the mycelium of the actinomyces is alone colored violet, when it has been longer exposed mycelium and conidia are stained alike.

Preparations of some value may also be obtained by the employment of picro-carmine alone; the fungi are stained yellow, whilst the neoplasm is red, but by this process we do not obtain a definition of the structure of the parasite.

MM. Cornil and Babès recommend double staining by the method of Gram with eosin or safranin. They thus obtain a violet staining of the central filaments whilst the clubs or conidia are stained red.

Cultures.—Bostroem obtained cultures upon beef blood-serum and upon agar, and has been able to convince himself that the bright claviform enlargements of the periphery of the colonies are incapable of multiplication, this property belonging to the central filaments alone; hence, he proposed classing the actinomyces with the schizomycetes or bacteria. The structure would be that of a cladothrix or branching bacterium, and the swollen productions of the cortical zone only incidental forms developing when the parasite finds itself under special conditions of nutrition.

We have cultivated the actinomyces in alkaline bouillons and our researches have led us to the same conclusions. Inoculation of peptonized glycerin-bouillon with a particle of tumor produces a development already well marked at the end of the first day; at this time may be seen a viscid, coherent layer floating near the bottom of the nutrient fluid which is still perfectly limpid, although holding in suspension a few white or yellowish, irregularly shaped grains. In examining these last under the microscope, they are found to be composed of ramified filaments, a few of which bear one or more distinct enlargements in every respect similar to the conidia of the cortical zone of the actinomyces. This structure is very evident, the mycelial filaments being yet loosely intermingled. Each of the latter consists of a tube containing, from point to point, a rounded nucleus which has a well marked affinity for the aniline colors. The bouillon acquires a fetid odor.

On the following days the pullulation continues with great activity, but the culture changes its appearance after the third day; the slimy growth in the bulbs resolves itself into a cloud of very fine granules which fall to the bottom of the bouillon as if the mucoid substance which held them suspended had disappeared; at the same time, the two layers of the liquid, up till now clearly defined, spontaneously mix themselves.

In examining bouillon cultures with the microscope, after the third day, the claviform enlargements are no longer to be seen. The small white granules deposited at the bottom of the bulbs are found to be composed solely of fine filaments, presenting lateral ramifications more abundantly than in the preceding preparations, and showing always in their interior the rounded nuclei, placed at regular intervals; these mycelial tubes are united in felted balls from which some filaments separate themselves to extend outward, or they form small tufts the branches of which radiate around a certain point in one of these filaments. Besides these groups of filaments a few may be observed which seem detached, but which have the same structure and are also branched.

We have succeeded in obtaining several generations of these cultures, starting from a tumor of the ox. We have also made cultures with the products obtained by inoculating the disease to rabbits. In these last cases the results have been the same, except that we have never seen any thing resembling the conidia.

Cultures are much more active when protected from the air; the actinomyces is, therefore, chiefly anaërobic. Various investigators have cultivated the fungus upon potato; the surface of the latter becomes excavated during the first days, then covered with colorless colonies with irregular surface, which soon becomes prominent and powdery; they become gray in color, then yellow, or even greenish when exposed to the light. Domec, in studying these cultures, has recently come to the conclusion that the actinomyces is a mucedine.

Experimental inoculations.—Johne has transmitted the disease to the calf and to the cow by subcutaneous, intra-peritoneal, and mammary inoculation; Ponfick and Israël have communicated it to the calf and to the rabbit. We, also, have inoculated actinomycosis to the rabbit both with artificial cultures and with the natural products obtained from a cow. The lesions remained local in all cases, and consisted of an exhausting suppuration which involved the death of the subject in from seven to ten days.

Etiology and pathogeny.—The actinomyces infects only the herbivora and omnivora; hence, the disease has been attributed to the forage. Johne found an identical fungus on the surface of husks of barley arrested in the tonsils of a healthy pig. Piana discovered, in a tumor of the tongue, vegetable debris along with the actinomyces. Most writers agree in incriminating more especially straw and barley husks.

The parasites, distributed on the ground by diseased animals readily bring about the transmission of the disease to healthy animals; Stiénon records an instance of enzoötic actinomycosis in which nearly all the cattle of the farm were affected.

Inoculation has also been successfully performed by Johne. This author injected under the skin and within the abdomen of two calves, and into the udder of a cow, the juice of a tumor of the maxilla and by this means produced the characteristic neoplasms. Similar results have been obtained by Ponfick and Israël in the calf and in the rabbit. This last author made his inoculations with the actinomycosis of man; the effects obtained were the same as with the tumors of cattle. Finally, accidental contamination has been established in persons who have attended to animals affected with the disease.

The lesions being most frequently confined to some portion of the digestive canal, it is logical to infer that the natural infection occurs through accidental solutions of continuity of the mucosa of these passages, principally in the anterior passages (wounds of the gums, carious teeth, crypts of the tonsils, etc.)

But penetration may also probably take place through the respiratory tract, the germs being transported by the air, by foreign bodies, or merely by the mucus.

Cutaneous wounds are also channels of introduction of the parasite (actinomycosis of the testicular cord in the horse; actinomycosis of the leg in the same animal in consequence of a kick; subcutaneous abscesses in cattle).

The tumors of actinomycosis (*actinomycomata*) result from the irritation excited by the actinomyces; they frequently have the structure of granulomata; the parasite nearly always occupies the center of an inflammatory nodule which may readily be differentiated into two zones, an internal, formed of epithelioid cells, and a peripheral, containing fusiform and lymphoid cells and indicating the transition to the condensed connective tissue which forms the frame work of the tumor. The fungus may also very frequently be seen surrounded by a row of giant cells. This structure resembles that of the tubercle type and shows that the pathogenic action of the actinomyces consists in a circumscribed and chronic irritation of the tissues into which it has been carried. After these tumors become softened in the center they contain a puriform yellowish or white liquid in which are suspended yellow grains, most frequently hard and calcified; these grains are the actinomyces. When the softening is regular the tumor acquires a characteristic spongy appearance.

It may happen, and this is the rule in man, that the actinomycotic tumors become complicated with suppuration. It appears to be established that the formation of pus should be attributed to the actinomyces themselves and not to pyogenic bacteria associated with them. Without excluding in all cases the intervention of the latter we think that the presence or absence of suppuration may be dependent upon the animal species infected, as well as on the tissue invaded. Thus, the injection of pure cultures of actinomyces within the abdomen of the rabbit always occasions a suppurative peritonitis; in man, also, the tendency to suppuration is always very marked. Stiénon records an instance of cattle affected with subcutaneous abscesses of an actinomycotic nature.

The tissue of the organ in which the tumor develops shows alterations corresponding to the chronic inflammations set up by its presence; certain parts become hypertrophied, whilst others diminish or may even disappear. Thus are explained the fibrous induration of some of these tumors, the osseous stalactites which develop on the maxilla when the invasion begins in the periosteum, and the cavities into which the same bone becomes hollowed when the pathological process takes place in its center.

Lesions may appear in the neighborhood of the first affected part; they result from the progressive enlargement of the primary tumor, or from a vascular emigration of the fungus.

The gradual increase in size of the neoplasm may lead to its extension to the skin or may give rise to distension, thinning, and perforation of the latter in one or more points, from which protrude fungus-like growths which ulcerate and secrete pus mixed with the yellow, pathognomonic grains.

The lymphatic vessels may convey the parasite to the corresponding glands and the latter then become the seat of new tumors.

The blood vessels sometimes transport the fungus, as has been established in man, in consequence of the ulceration of the jugular vein in contact with a diseased focus.

Generalization does not appear to take place in animals, perhaps on account of the premature slaughter of the subjects, a course which is most frequently followed and which is the most economical. Actinomycosis has, however, been encountered in the spleen in cattle and in the muscles in pigs.*

Besides the functional disturbances to which they give rise (disturbances of mastication, deglutition, respiration, etc.) the actinomycomata, especially when they have ulcerated, cause a progressive emaciation of the diseased animals.

Botryomycosis.

In certain indurations of the testicular cord of the castrated horse a special parasite is found to which the various names of *botryomyces* (Bollinger), *discomyces* (Rivolta), *botryococcus ascoformans* (Kitt), etc., have been given. It has also been met with in certain forms of fistulous withers, in tumors at the point of the shoulder, and in certain nodosities of the skin and subcutaneous tissue, etc. Czoker has also re-

* [The actinomyces-like parasite found in the muscles of swine (Actinomyces musculorum suis) is not identical with the Actinomyces bovis *sive* hominis. C. Günther: *Bakteriologie, p.* 327.—D.]

corded its occurrence in the pus of a chronic mammitis in the cow.

The parasite appears in the form of a cluster of small spheres full of micrococci. The latter are united by a gelatinous substance, and each of the spheres is inclosed by a double contoured membrane. New spheres develop at the periphery of the colony. The parasite is related to the genus ascococcus.

Fig. 15.

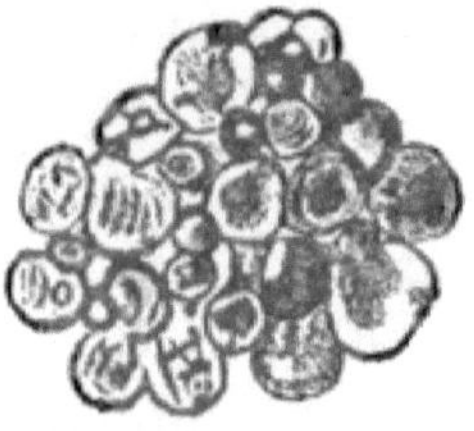

Botryococcus ascoformans. Slightly magnified. (Kitt.)

Staining may be obtained by the method of Gram, with eosin as contrast stain. The eosin becomes fixed especially on the gelatinous fundamental substance; picric acid has a similar action.

The pathogenesis of the lesions (botryomycomata, mycofibromata, mycodesmoides) is the same as that of the actinomycomata. They are inflammatory tumors of slow evolution showing a mixture of granulation tissue and fibrous tissue of different degrees of condensation. The granulation tissue is often arranged in tuberculiform masses of which the parasite occupies the center, and the elements of which may suppurate, become necrosed, and give rise to fistulous tracks.

The parasite generally confines its ravages to its primary seat; it may, however, migrate toward the lymphatic glands and may even pass into the blood. Metastatic foci then appear, commonly localized in the lung, skin, etc. In the lung these lesions have a resemblance to those of glanders.

The germ gains entrance to the system through operative (castration) or accidental wounds.

Bovine farcy.

This disease is peculiar to the bovine species. Formerly noticed in France, it appears to be much rarer at the present time; it exists in the Guadalope. It is usually located on the limbs, manifesting itself by painless cords along the course of the subcutaneous veins and terminating at the corresponding lymphatic glands; the brachial, prescapular and prepectoral glands are those most frequently attacked. These cords and glands may suppurate; the suppuration is always slowly evolved and the whole assumes the characters of a cold abscess, with very thick and indurated walls. After these abscesses are opened the subject seems to recover, but very soon other tumors appear. The animal pines away and dies by slow decline. The autopsy brings to light pseudo-tuberculous lesions with purulent centers in the internal organs: lungs, liver, spleen, and lymphatic glands.

We are indebted solely to M. Nocard for the bacterial study of the disease.

Microbe.—There is found in the pus of the abscesses and in the center of the pseudo-tubercular lesions "a long, slender bacillus, appearing under the form of small, intricately interwoven masses, the central part forming an opaque nucleus from which radiate to the periphery myriads of fine prolongations, the majority of which seem to branch." The dimensions of this bacillus are nearly the same as those of the bacillus of rouget.

Fig. 16.

Bovine farcy.
(M. and L.)

The microbe of cattle farcy is purely aërobic.

Cultures.—Cultures are easily obtained between 30° and 40°. In bouillon it forms rounded pellicles of a dull gray color and oily appearance, floating in the liquid. On agar and gelatin it forms small masses more or less regularly rounded, opaque, and thicker at the periphery than at the center. On potato the growth is rapid and takes the form of dry salient plates, often depressed at their center.

The microbe reproduces itself in cultures in the form of star-like masses of intertwining filaments, and it is only at the periphery that the bacillar nature of its constituent elements can be recognized. According to M. Nocard, the ramifications observed in these masses are false; the bacillus divides transversely, then the terminal filament inclines to a right angle, allowing its generator to continue its direct course. The organism, therefore, is not a true cladothrix.

Research and coloration.—The bacillus of cattle farcy is easily colored; it is decolorized by the reaction of Gram when the contact with the alcohol is too prolonged, but it takes the double stain of Weigert perfectly. The spores are stained with difficulty.

Experimental inoculations.—The disease is inoculable to cattle, to the sheep and to the guinea pig; not to the rabbit, cat, dog, nor horse.

As a result of intra-peritoneal inoculation in the guinea pig the serous lining becomes studded with tuberculiform nodules, in the purulent center of which the characteristic tufted bacilli are contained; these nodules are especially abundant in the omentum, which itself has become much enlarged. The abdominal viscera are altered only in their serous covering, the tissue proper being unaffected, whilst the thoracic organs are exempt from lesions.

The injection of virulent material into the veins is followed by a generalized pseudo-tuberculosis, the bacilli being found in tufts in the center of all of the lesions.

In both cases death occurs from the ninth to the twentieth day. Subcutaneous inoculation in the guinea pig always produces a voluminous local abscess; the corresponding lymphatic glands also suppurate and recovery takes place only after a period of great emaciation, leaving behind it an induration of the lymphatics. Generalization may, however, be occasionally observed.

Intra-vascular inoculation in the sheep and in cattle also produces nodular lesions distributed throughout the viscera, but death does not immediately follow, the disease developing very slowly.

Hypodermic inoculation in the same animals produces an abscess which ulcerates on different occasions, seems to heal, then reappears later, exactly as in the natural disease.

Subcutaneous inoculation in refractory animals occasions an abscess which quickly heals.

Tetanus.

Tetanus occurs spontaneously in all the domesticated animals and in man; it is more frequent in the horse, ox, sheep, and goat, but it has also been observed in the pig and dog.

The contagiousnesss of tetanus, which had long been suspected from clinical observation, has been demonstrated by experimental inoculation, and its causative agent has been brought to light by bacteriological research.

Microbe.—The bacillus of tetanus, first described by Nicolaier, is a slender rod, measuring 3μ to 5μ in length, homogeneous, or containing at one end an enlargement in which appears a spherical spore of a diameter double or quadruple that of the filament. It has, therefore, when sporulated, the form of a pin, of which the brilliant, lustrous spore represents the head. In wounds the bacillus sometimes attains a length much greater than that indicated above.

Fig. 17.

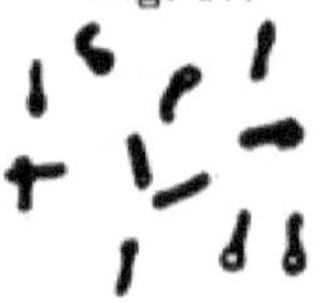

Bacillus of tetanus, sporulated and non-sporulated. (M. and L.)

The bacillus of Nicolaier is motile, its movements resembling those of the septic vibrio, but these movements cease when fructification is accomplished. It is anaërobic, multiplying only when the atmospheric air is excluded. A rarefied atmosphere is, however, compatible with its vitality, and this will account for its mutlipication in free media, the aërobic germs protecting it from contact with too large a proportion of oxygen.

The tetanus bacillus remains intrenched in the infected focus and does not penetrate into the blood during life; at the approach of death, or a certain time after death, it may be found in remote parts in which the deoxygenation of the blood has allowed of its multiplication.

The bacillus is met with in its two forms, homogeneous and sporulated, in the pus of wounds which have given rise to tetanus, the sporulated form, however, being less abundant than the non-sporulated.

In young cultures the latter form also predominates.

Action of physical and chemical agents.—The spores are very resistant to heat; they support a temperature of 80° during six hours, or 90° during one hour. In moist heat at 100° they are killed in a quarter of an hour, at 115° in five minutes. Spores which are thoroughly dried retain their vitality for a long time if protected from the light. Exposed to the light they do not survive more than one month especially if, at the same time, exposed to the air.

The bacilli of Nicolaier are little sensitive to the action of antiseptics, notably less so than the septic vibrio. Thus, sublimate solution, at 1 to 1000, only kills these bacilli after three hours; three per cent carbolic acid solution after ten hours. The spores resist a five per cent solution of carbolic acid for fifteen hours, the same solution with an addition of 0·5 per cent of hydrochloric acid, two hours, and 1 to 1000 sublimate solution with a like proportion of hydrochloric acid, thirty minutes. They are unaffected by the gastric juice and by putrefaction; active cultures having been administered to rabbits, guinea pigs, mice and dogs by way of the mouth, all these animals remained healthy, whilst their excrements were virulent for other animals.

Cultures.—The bacillus of Nicolaier vegetates in artificial culture media when the oxygen of the air is excluded and when under suitable conditions of temperature; culture vessels should therefore be employed in which a vacuum has been created or in which the atmosphere has been replaced by hydrogen, carbonic acid, etc. Solid liquefiable media, such as gelatin and agar, are sometimes freed from air by boiling for half an hour; the mass is solidified by rapid cooling, inoculated by deep puncture, and the surface of the medium covered with a layer of sterilized oil which prevents the absorption of the atmospheric oxygen. Deoxygenation can also be accomplished by the addition to the same media of substances which readily absorb oxygen. Kitasato advises the addition to gelatin or agar, of glucose two per cent, sulpho-indigotate of soda 0·1 per cent, blue turnsol five per cent.

The most favorable temperature varies between 38° and 39°, but multiplication still occurs at 18° and continues even at 43°.

The natural inoculation substances at our disposal: pus of wounds which have given rise to tetanus, earth, forage, etc., are always contaminated with other germs than those of this disease; hence, impure cultures are first obtained from which the bacillus of Nicolaier must be isolated. With this aim we take advantage of the great resistance of its spores, and heat the sporulated cultures in a water bath at 80° to 90° for three quarters of an hour to one hour; traces of these cultures are sowed in anaërobic tubes which are then closed over the flame and the gelatin spread out on their walls by a rotatory motion.

The colonies which then develop are composed exclusively of tetanus bacilli, as may be proved by inoculation. If some spores of the septic vibrio have resisted the heat and produced a collateral vegetation, we should then resort to the action of antiseptics; 1 to 1000 sublimate solution kills the tetanus bacillus only after three hours, 3 per cent carbolic solution only after two hours. (Sanchez-Toledo.)

In bouillon the development is very rapid; the liquid becomes turbid in one day and sets free small bubbles of gas; the growth abates and becomes deposited toward the fifteenth day.

When a gelatin tube is inoculated by deep puncture and kept at 18°, the growth forms, after four or five days, small cloud-like points from which fine lines radiate perpendicular to the puncture. The culture has a flocculent appearance, slowly fluidifies the gelatin, and bubbles of gas are disengaged; when the gelatin is entirely fluidified the culture is deposited in the form of white flakes.

Cultures upon agar are less characteristic. Growth also takes place upon serum and potato. Cultures of tetanus emit a smell of burnt horn and produce various gases, including carbonic acid and hydrocarbons.

Research and coloration.—The bacillus of tetanus is easily stained with the aniline colors; it takes the Gram very well and appears as a slender rod, uniform in size or swollen at one of its extremities. Before the formation of the spore this enlargement stains like the filament itself. The spores are not stained by this process but may be colored by the method usually employed for the staining of spores.

Experimental inoculations.—The disease is inoculable to the smaller animals and notably, with a susceptibility decreasing in the order in which they are named, to the mouse, white rat, guinea pig, rabbit, dog, pigeon and chicken. The mouse, white rat and guinea pig are extremely susceptible, 0.002 cub. cent. being sufficient to produce in these animals the typical disease with fatal results in 36 to 40 hours. The rabbit requires 10 to 30 drops; the symptoms appear from the second to the third day and death occurs four to ten days later. The dog, pigeon and chicken are less susceptible and sometimes survive large doses.

The inoculation succeeds well in the connective tissue, in the peritoneum and in the arachnoid. In the connective tissue it causes an œdematous swelling, the contractions appearing first of all in the adjacent muscles. When inoculated in the peritoneum or in the blood these appear simultaneously in all parts of the body. The disease is not produced either by inhalation or ingestion of virulent products. Ingestion of tetanic toxines is also without effect, these toxines being destroyed by the digestive juices.

According to the dose inoculated and the susceptibility of the animals the experimental disease may be acute or chronic, fatal or curable.

In connection with experimental inoculations three cases must be distinguished. The material to be injected may consist of the entire culture, of the amorphous part alone freed from its microbes by filtration, or, finally, of the microbes alone deprived of their soluble products by filtration, lixiviation, or heat.

In the first two cases—inoculation of the entire culture or of its soluble products—the classic disease is reproduced. In the last case, according to MM. Vaillard and Vincent, the inoculation remains without effect except when excessive doses are employed. The tetanus bacillus, therefore, appears to be unable to multiply in the organism in the absence of its toxines, being destroyed, in such case, by the phagocytes; it excites an active diapedesis which may be observed after injections into the anterior chamber of the eye. The toxines of cultures, on the other hand, seem to repel the leucocytes and thus protect the microbes from their destructive action. Certain other substances may take the place of this action: inoculation of spores alone is followed by tetanus when lactic acid, diluted to 1 to 500, or trimethylamin, is injected at the same time, or when a contusion is produced in the inoculated tissues. Simultaneous injection of the germs of tetanus and the micro-bacillus prodigiosus also allows the irruption of the disease; the prodigiosus attracts the phagocytes to itself and thus forms a barrier behind which the tetanus bacillus multiplies and secretes its protective toxines.

The experiments of Vaillard and Vincent have been repeated by Sanchez-Toledo, who seems to have succeeded in transmitting the disease by means of cultures freed from their toxines and from all adjuvant substances.

Etiology and pathogeny.—The tetanus bacilli or their spores exist in the soil; the disease has often been produced by inoculation of water with which certain specimens of soil have been washed. From the soil they are transported in the dried forage, pass unchanged through the digestive canal of herbivora and with the manure are returned to the soil. By inoculations they have been demonstrated in hay and in the excrements of healthy horses and cattle. Tetanus is therefore of telluric origin; but it can also be transmitted from one animal to another, either directly or indirectly. The microbe being inactive when inhaled or ingested, it must gain entrance to the system through a solution of continuity.

All wounds are not alike adapted to the growth of the tetanus bacillus; the germ is anaërobic and consequently requires a medium little accessible to atmospheric air; besides, it is inoffensive if not protected against the phagocytes by accessory conditions. The production of tetanus, therefore, as a rule, requires deep, anfractuous, contused wounds or those contaminated by other germs, and more especially by the ordinary pyogenic species. Nevertheless, tetanus has often been observed to follow insignificant wounds. It should be observed that wounds favorable for the growth of the bacillus of tetanus are also favorable for that of the septic vibrio. Now, the bacillus of malignant œdema is also found in tetanogenic earth, so that the same wound may be contaminated with both germs; gangrene, however, running its course more rapidly than tetanus, the latter may only appear after recovery from the former, or may not appear at all if the subject succumbs to the septicæmia (Verneuil).

The bacillus of tetanus secretes in the wound special toxines which have a poisonous action on the organism similar to that of strychnine. If a culture of tetanus be filtered so as to completely separate the microbes from the soluble part, and the latter be injected to animals, an absolutely typical tetanus results. This filtered liquid is very toxic: one-twentieth of a drop kills a mouse in thirty-six hours, one drop kills a guinea pig in twenty-four hours.

The nature of the tetanic poison is yet incompletely known in spite of numerous investigations.

Brieger isolated from impure cultures on meat several toxic ptomaines: *tetanin*, *tetanotoxin* and *spasmatoxin*. The first gives the typical tetanus when injected in very small doses, the second produces tonic and clonic convulsions, and the third hyper-salivation and convulsions. According to more recent investigations the tetanic poison appears to be an albumin related to the diastases or soluble ferments. It is destroyed by a temperature of 65° in five minutes, is insoluble in alcohol, soluble in water, and adheres to precipitates of alumina and phosphates. It is destroyed by heat and preserved by desiccation if this is rapidly obtained in a vacuum and without the aid of heat.

This toxine appears to act more particularly on the muscular tissue, which would explain the appearance of the first tetanic contractions in the muscles which are adjacent to the wound or to the place of inoculation. When administered by way of the digestive canal it is inactive.

From the preceding considerations it may be inferred that the microbe of Nicolaier produces its effects only by means of the diastases which it secretes. Moreover, it gains entrance into the blood only in the last moments of life, or after death.

Attenuation. Vaccinations.—By heating the filtrate to different temperatures between 55° and 100° until it becomes inactive, and inoculating this material to mice, we determine a mild, curable form of tetanus, but one which is not followed by immunity.

Kitasato endeavored to obtain immunity against tetanus by the action of trichloride of iodine. He injected 0·3 cc. of filtered culture under the skin of a rabbit and immediately afterward, in the same place,

3 cc. of 1 per cent solution of trichloride of iodine, the latter injection being repeated every twenty-four hours until the rabbit had received in all 0·15 grams of the trichloride. If, after fourteen, eighteen, twenty-five days, 2 cc. of the virulent filtrate, or 2 to 3 cc. of virulent bouillon culture, be injected, the tetanic symptoms which supervene disappear in a few days and the animal can then receive an injection of 5 cc. of virulent culture without manifesting any symptoms. The rabbit is therefore vaccinated; but this process is inconstant, and sixty per cent of the rabbits do not obtain immunity. The mouse and guinea pig do not obtain it at all. Now, the serum and blood of these vaccinated rabbits possess the very interesting property of destroying the toxine; they are toxinicidal. A mixture, twenty-four hours old, of 1 cc. of a virulent culture with 5 cc. of this serum can be injected with impunity to mice. It is even possible by means of this serum to check the disease experimentally developed in the mouse. Further, the injection of 0·2 to 0·5 cc. of this serum into the peritoneum of the mouse confers an immunity of from forty to fifty days' duration.

Tizzoni and Cattani have obtained immunity in the dog and pigeon by the injection of progressive doses of cultures of gradually increasing virulence. The serum of the dog, thus vaccinated, is toxinicidal and confers immunity on the dog and on the mouse but not on the rabbit or guinea pig, and does not check the disease when already established. These authors have isolated and obtained in the dry state the toxinicidal product of the serum (*antitoxin*); they appear to have successfully used it in the treatment of human tetanus.

Vaillard, by injecting into the blood of rabbits, first, in several doses, 40 cc. of filtered culture heated to 58°, second, 10 cc. of filtered culture heated to 51°, and, finally, 15 cc. of filtered culture not heated, did not immunize these animals, but communicated to their blood the toxinicidal power; hence it follows that there is no relation between the toxinicidal property and the refractory state.*

* [Of the more recent investigations bearing on the etiology of tetanus and the production of immunity, the following points may be briefly noted:

The necessary co-operation of other microbes at the place of infection (in the naturally acquired disease) has been reaffirmed. (Vaillard and Rouget.)

According to Courmont and Doyon, the special tetanus toxine results from a fermentation excited in certain tissues of the organism by a soluble ferment secreted by the bacillus tetani. (*Compt. rend. Soc. de Biologie*, March, 1893.)

Immunity can be obtained by repeated injection of cultures of the bacillus tetani grown in bouillon prepared from the thymus gland of calves, or of filtered cultures to which a certain proportion of an extract of this gland has been added (Brieger, Kitasato, etc.); the blood serum of animals thus immunized possesses immunizing properties. By repeated injection of gradually increasing doses of virulent cultures to animals (rabbits, dogs, horses) which have been thus immunized, the protective power of the serum of these animals becomes greatly increased. This serum then furnishes an effective "vaccine." The serum of such immunized animals or a preparation from the same (anti-toxin: Tizzoni, Cattani) has been used (by repeated subcutaneous injection of considerable doses) in the treatment of tetanus in man, and apparently with good results. (Ref. *Centralbl. f. Bacteriologie*, *XIII*, 4, 14; *XIV*, 4, 12, 19; *XV*, 4.)

The curative action of the serum of immunized animals depends upon its immunizing properties; it localizes the tetanus

Diphtheria.

In its widest meaning this term is applied to a special form of inflammation of the integuments in which a concrete exudate is produced in the thickness of the derm and involves its mortification; this exudate is called *diphtheritic.* When the fibrinous deposit is limited to the epithelium the inflammation is said to be *croupous.* These two forms of inflammation can be produced by very varied causes: these are mechanical (compression), or physical (burns), or chemical (caustics), or, finally, biological (parasites and microbes). Among the number of parasites may be mentioned the gregarinæ of the contagious epithelioma of poultry and the coccidia of typhlitis of the same animals; among microbes a large number possess the same faculty; of these we need only recall the diphtheritic exudates of pneumo-enteritis of the pig, of acute glanders, and of petechial typhus, etc. *Diphtheria of wounds*, or hospital gangrene, characterized by a superficial necrosis of the divided tissues, must also be attributed to the contamination of the latter by micro-organisms.

From a clinical point of view the term *diphtheria* refers to a specific disease due to a special germ, and manifesting itself by an inflammation, with croupous or diphtheritic evolution, of the respiratory and some-

by protecting the parts of the nervous system not yet attacked by the tetanus poison. (Tizzoni, Cattani: *Id. XV*, 17.)

Tests of the curative action of the serum of immunized animals made by Nocord, in the case of two sheep artificially infected with tetanus, resulted unfavorably. (Ref. *Jour. Comp. Path. VII*, 1.)—D.]

times digestive mucous membrane. This disease has been thoroughly studied in man; it also occurs in certain species of animals, particularly in birds and in the calf. At one time it was believed that human and avian diphtheria were identical, but this view is no longer entertained. The two diseases have a different evolution and different microbes.

Human diphtheria.—Klebs and Löffler discovered in the false membranes a bacillus, straight or curved, with rounded ends, sometimes club-shaped, at other times less than the average thickness. It measures 2·5μ to 3μ in length by 0·7μ in thickness; it is especially aërobic but also grows when the air is excluded; it is met with in the superficial zone of the false membranes. The germ grows on most of our artificial media; it is stained by Löffler's methylene blue and by the Gram method. Roux and Yersin have completed its biological study; they isolated it by inoculating a series of serum tubes without recharging the platinum wire: the colonies of diphtheritic bacilli appear as grayish-white rounded growths with the center more opaque than the periphery.

Excoriated mucous membranes, inoculated with a culture, soon show the characteristic false membrane in rabbits, guinea pigs, cats, pigeons, and chickens.

The subcutaneous injection of a few drops of the liquid kills the guinea pig in thirty-six hours with general vascular dilatation and pleural effusion. The rabbit requires 1 cc. of the culture and, like the pigeon, dies after several days. The dog and the sheep also succumb. In all cases a hemorrhagic œdema is produced at the place of inoculation and the vessels of the different organs are congested. When death of the inoculated animals is delayed they exhibit paralysis resembling that observed in the child suffering from this disease.

The bacilli occur only in the specific lesions; there they secret a poison the absorption of which determines the general symptoms of diphtheria. Cultures freed from microbes by filtration are very toxic; under their influence guinea pigs exhibit a pronounced dyspnœa; rabbits are attacked with progressive paralysis and often with diarrhœa. The dog and the sheep also succumb after showing symptons of paralysis.

The toxic substance of cultures is a diastase presenting much analogy with that of tetanus; in the digestive canal it is innocuous.

Behring appears to have obtained immunity in guinea pigs: 1st, by inoculating them with cultures sterilized at 65° to 70° C.; 2d, by injecting them with a mixture of one part of trichloride of iodine and five hundred parts of these same cultures; 3d, by inoculating them with the serous or sanguinolent liquid taken from the pleura of guinea pigs dead of diphtheria.

Avian diphtheria.—Birds are very subject to a disease manifesting itself by the production of false membranes and diphtheritic exudates on the mucous membranes of the mouth, pharynx, œsophagus, nose, eyes, larynx, trachea, lungs, air cavities, intestines, and upon the skin. This disease is very contagious although much less severe than that of man; like the latter it may occasion rapid mechanical asphyxia, but more frequently it is protracted, presenting remissions and exacerbations, and leading to the emaciation of the affected birds. A catarrhal inflammation which sometimes occurs in the attacked mucous membranes, by provoking glairy morbid secretions, aggravates the cachexia and hastens the fatal termination.

Löffler attributes the disease to a special bacillus, staining by the methods of Gram and Weigert, which he met with in the fluid products of the inflamed mucous membranes, in the false membranes, the lesions of the liver, and in the blood. This bacillus has nearly the same dimensions as that of human diphtheria but it is smoother and more uniform in thickness. It is abundantly present in the superficial layer of the false membranes, rare or absent in the deep layers. Along with this specific bacillus other microbes are constantly found in the concrete exudates.

Löffler reproduced avian diphtheria in the rabbit and the pigeon; death, when it occurred, supervened less quickly than with the human bacillus. The guinea pig and the dog are also more sensitive to the latter bacillus than to that of birds. Löffler cultivated and described the microbe and with it reproduced the disease in birds and the rabbit by inoculation of cultures on the mucous membrane, on the skin, and in the subcutaneous cellular tissue.

The disease is propagated by means of the morbid matters (discharge, fæces, false membranes) of the diseased animals. The virus gains entrance into the organism of healthy animals with the food or by the air. Contamination most frequently takes places in an indirect manner but may also occur directly. M.

Megnin has recorded an outbreak in pigeons in which the young animals quickly succumbed to diphtheria; he attributes the disease of the young pigeons to the fact that the mothers were affected with old, little marked œsophageal lesions the virulent products of which mingled with the lactescent liquid of the crop were fed directly to the young.

The disease appears in poultry pens in consequence of the importation of birds coming from infected centers or on the return of animals which have been exhibited in bird shows. The virus appears to be preserved for a considerable time in contaminated pens and in manure.

In spite of the cases advanced by different observers it appears to be well established that the disease of fowls is not transmissible to man.

Diphtheria of calves.—There exists in the bovine species a diphtheritic affection transmissible chiefly to calves and also having its origin in a special bacillus. It localizes itself in the back of the throat, in the trachea, and in the bronchi.

Rabies.

Rabies occurs, under natural conditions, in all domesticated animals and in several wild species: wolf, fox, jackal, bear, etc. It is very much the most frequent in carnivora, and more especially in the dog and wolf.

Microbe.—In despite of numerous researches concerning the pathogenic agent of rabies, its morphological characters are yet unknown. Its existence however can not be doubted; it multiplies in the organism to which it is inoculated, and loses or gains in virulence like the well defined microbes of other diseases. Fol and Babès have each described micrococci which they were able to cultivate in bouillon, and with which cultures they believed that they transmitted rabies. Babès also describes a short bacillus. Mottet and Protopopoff isolated from the brain very fine bacteria, cultures of which in bouillon, according to these authors, gave the typical disease.

Action of physical and chemical agents.—The virulence of an emulsion of the spinal cord of a rabid animal is dissipated when heated twenty-four hours at 45°, one hour at 50°, and half an hour in steam at 100°.

It resists a temperature of —20° for thirty hours at least. It is destroyed by fourteen hours exposure at 37° to solar light. Desiccation in the air rapidly diminishes its virulence and destroys it in a few days.

Virulence is retained in cadavers for several weeks if decomposition is prevented, but it is destroyed by putrefaction. Under the influence of 1 to 1000 sublimate solution, 2 or 5 per cent solution of permanganate of potash, or 50 per cent alcohol, the virus is quickly impaired; 15 per cent alcohol, on the other hand, preserves it intact for seven days at least. Even a large dose of the emulsion proves inoffensive when it has been rendered either acid or alkaline. Perfectly neutral glycerin (30 B) preserves its pathogenic power in a perfect manner.

Experimental inoculations.—With the exception of some individuals which are naturally insusceptible, the majority of the mammalia contract rabies when inoculated.

The species used in the laboratories are the rabbit, guinea pig, dog, ape, rat, and birds. The last named species almost invariably recover spontaneously (Gibier). The virulent material is taken from the nervous substance (brain, medulla, cord and nerves), from the salivary glands, the saliva, bronchial mucus, and from the pancreas. The milk occasionally shows itself virulent, but the blood is not; complete transfusion of blood from a rabid dog to a healthy subject did not produce the disease (Bert.) In practice we resort more especially to the nerve centers, which furnish a pure virus; for use, a particle is reduced to pulp and the latter diluted in sterilized water or bouillon; in this way a white milky fluid is obtained, a veritable emulsion of the nervous substance.

The inoculation may be performed in various ways; it may be subcutaneous, intra-muscular, intra-venous, intra-ocular, intra-cranial, intra-nervous, etc., the result varying according to the method employed.

In the dog, subcutaneous and intra-venous inoculations are usually followed by dumb rabies, without barking or fury. Furious rabies can be obtained by the same methods, but with small doses of virus. The smaller the amount of the virus employed the more readily is the furious form of the disease obtained, the period of incubation being at the same time prolonged. The injection of very small doses may be ineffective in the dog without conferring immunity, whilst inoculation of large quantities of virus may give immunity without the manifestation of any symptoms of rabies.

Intra-muscular inoculation is followed by rabies with more certainty than subcutaneous inoculation.

In the dog, intra-cranial inoculation is always followed by furious rabies; the period of incubation for the virus of the streets is from fourteen to fifteen days, on an average.

The inoculation may also be made in the nerve trunks, a previous lesion of the nerve fibers increasing the chances of success; in this method the period of incubation is not the same in all cases, and is always longer than by trephining; the virus vegetates in the nerve and progresses from the periphery toward the center, the disease being more tardy in its appearance as the course to be traversed by the virus is more extended. Inoculation in the posterior nerve trunks is followed by paralytic rabies; in the anterior trunks by furious rabies.

The insertion of the virus into the anterior chamber of the eye is invariably followed by rabies, the incubation in this case being thirteen to sixteen days for the natural virus of dogs.

In the rabbit, intra-cranial inoculation enables us to make some important observations: the incubation is from fifteen to seventeen days, and death occurs in the course of the next four days. The disease manifests itself by paralytic phenomena with progressive course; rabies of the rabbit is, therefore, dumb rabies. However, cases of furious rabies are also observed, and Ferré has shown that symptoms of excitement (accelerated respiration, etc.) precede the paralytic symptoms.

Passage of the virus of natural rabies from the first rabbit to a second, from the second to a third, and so on, exalts the activity of the rabic virus. This exaltation shows itself in the shortening of the period of incubation which becomes abbreviated to eight days after twenty-five passages and to seven days after fifty passages. The virus is then acclimated in the rabbit, which dies regularly after seven days whatever generation the virus employed may belong to; the virus of rabies thus exalted takes the name of *fixed virus.* It is exalted not only for the rabbit but also for the dog itself, which always contracts the disease from intra-vascular inoculation of this virus, whilst similar inoculation of natural virus is uncertain in its results.

In the guinea pig, which readily takes rabies by intra-cranial inoculation of natural virus, virulence is exalted by passage in series, until, after the eighth inoculation, it becomes fixed. The incubation period is then five days. Rabies in the guinea pig often shows a distinct period of excitement; the fixed virus of the guinea pig is more active for the dog than the natural virus.

In the ape, passage in series attenuates the virulence; whilst the virus taken from the first ape kills the rabbit in thirteen to sixteen days, that of the second allows an incubation in the rabbit of fourteen to twenty days, and that of the sixth, thirty days. Intra-venous inoculation of the virus of the sixth generation no more produces the disease in the dog, and even intra-cranial inoculation is uncertain. The virulence attenuated in the ape can be restored to its original activity by a series of passages through rabbits.

Etiology and pathogeny.—Rabies is undoubtedly caused by a living virus. It is transmitted from one animal to another by direct contagion. Mediate transmission seems to be of very rare occurrence. The virus is preserved only for a very short time in external media, a circumstance which diminishes the chances of its transportation; M. Galtier, however, has pointed out the possibility of infection by the intact ocular mucosa, and it is quite conceivable that this might actually occur in man through the projection or transference of a virulent liquid to the eye. Contamination by ingestion of the flesh coming from a diseased animal appears to have occurred a certain number of times in the dog (Zundell). Experimental tests of this mode of infection have, however, given only negative results.

The disease is transmitted to the healthy individual in nearly all cases by the bite of a rabid animal, such bite being equivalent to an inoculation with virulent saliva. Bites of carnivora (dog, cat, wolf) are the most dangerous, on account of the severity of the wounds which they inflict; those of the horse and deer are less dangerous; bovines have not as yet been known to communicate the disease. MM. Nocard and Roux have shown that the virulence appears in the saliva, on an average, twenty-four hours before the first symptoms, and that there are cases in which it appears one or two days earlier. On this account they recommend that all dogs which have bitten other animals or human beings should be kept under observation for three days at least before certifying to their condition.

Bites which involve the nerves or muscles are among those which are the most likely to communicate rabies.

Besides bites, direct inoculation may occur in the

course of an autopsy. Inter-placentary transmission has not been established; cases of such transmission which have been brought forward have not been verified.

The period of incubation is very variable; the disease most frequently appears in the course of the first two months after the bite, but it may appear after a few days as well as after several months. These differences depend upon the more or less easy propagation of the virus. Absorption takes place chiefly by the nerves; in fact, inoculation in a nerve trunk induces rabies more rapidly than its insertion in the connective tissue at a corresponding point, although less quickly than when introduced into the cranial cavity. The virus, therefore, seems to vegetate in the nerve and progress toward the cerebro-spinal center, thence radiating along the nerves.

After inoculation in the sciatic the primary rabic symptoms indicate changes in the lumbar cord, the cervical cord and the medula being only attacked at a later stage. In one case in the guinea pig the disease was limited to paraplegic symptoms alone; by interrupting the continuity of the spinal cord at the level of the dorso-lumbar region the dorso-cervical cord was protected from the action of the virus. Some cases in man which have been carefully observed support to some extent this view as to the propagation of the virus of rabies: after a bite on the right arm the nerves of this arm, inoculated along with those of the left arm, alone showed themselves virulent. Incubation should therefore be shorter in proportion as the length of nerve to be traversed before arriving at the center is itself shorter. However, under natural conditions the duration of this latent period is influenced by various factors, such as the quantity of the virus, the richness in nerves, and the extent of the injury to these nerves.

The action of the virus of rabies on the nerve centers and on the nerves is indirect; this virus does not act as a chemical agent would, strychnine, for example. The anatomical changes, little visible to the naked eye, appear under the microscope as inflammatory lesions at different stages of their evolution. These lesions have their seat principally in the nerves of the bitten part and in the corresponding part of the spinal cord. They consist of congestions and capillary hemorrhages, with infiltration of leucocytes; other lesions observed are limited foci of necrosis, diverse degenerations of the central nerve cells, transformation of the myeline, hypertrophy of the axis cylinders, etc.

The rabid symptoms are the expression of the neuritis and myelitis which successively develop. These changes manifest themselves by phenomena of excitement or of depression according as they are more or less advanced. Such phenomena are usually successive in occurrence, but may appear separately; hence the two forms, furious rabies and dumb or paralytic rabies. Most frequently they are combined, and mixed forms result, intermediate between the two preceding, thus establishing the unity of the disease. The symptomatic tableau will naturally vary according to the part of the cerebro-spinal axis first invaded, and therefore according to the place of inoculation.

The experiments of Pasteur, in which rabies was

transmitted by intra-venous inoculation, demonstrate the possibility of the absorption of the virus and its transmission by the circulatory fluids.

Attenuation and Vaccination.—The virus of rabies is not absolutely fixed; its passage through the organism of the rabbit exalts its virulence, as we have seen above, whilst its inoculation to the ape enfeebles it. It becomes acclimated in the ape after three generations and then requires an incubation of twenty-three days, the incubation for natural rabies being only eleven days. The fixed virus of the ape only rarely communicates the disease to the dog by intra-cranial inoculation, and intra-venous inoculation is always inoffensive; this virus can be used to vaccinate the dog.

Babès has shown that the virus of rabies introduced into the lymph sac of the frog becomes progressively attenuated until, after a certain time, it is capable of acting as a vaccine for the dog. The lymph of the frog removed from the body and mixed with an emulsion of a virulent cord also attenuates the activity of the latter.

Attenuation by desiccation.—A virulent rabic cord progressively loses its activity by desiccation in the air. This attenuation manifests itself by retarding the appearance of the disease,—increasing the duration of incubation; thus, after two days' desiccation the fixed virus of the rabbit has not changed, and produces the disease in rabbits after seven days; after three to five days the incubation lasts for eight days; after six days it is extended to fourteen days; after a desiccation of more than seven days the disease is no more communicated to the rabbit.

In order to obtain a regular and stable attenuation Pasteur starts with the cord of a rabbit which has been rendered rabid by means of fixed virus; he divides it into segments of two centimeters in length and these he suspends in a wide-mouthed liter bottle containing pieces of caustic potash; this bottle is placed in a chamber in which the temperature is maintained at 20° C.

Variations in the temperature lead to modifications in the attenuation; at 28° the virulence is destroyed in five days; at 35° in four hours.

The Russian rabbit, which has a smaller cord, allows of a more rapid attenuation.

The temperature and the oxygen of the air are instrumental in producing this loss of pathogenic activity; the attenuation is slower in carbonic acid; the loss of humidity seems to have less influence; Protopopoff has, indeed, shown that this modication occurs also in glycerin-bouillon.

Authorities are far from being unanimous as to the theory of this attenuation. Pasteur thinks that it consists merely in an impoverishment in active virus and not in a diminution of activity.

The hypothesis of a chemical vaccine has been dispelled by Babès who showed that the rabic substance filtered through porcelain, as well as that heated to 100°, has no vaccinal property. Gamaleia has demonstrated that vaccinal injections have no action on an individual already affected with the disease, which also dispels the theory of a chemical vaccine. Virus attenuated by desiccation regains the activity of the fixed virus on its second passage through the rabbit.

Anti-rabic vaccinations.—It is possible to confer immunity against rabies both upon man and susceptible animals.

This immunity has been noted in certain cases of natural rabies in which recovery took place. It was especially well observed by Hoegyes in cases of experimental rabies followed by recovery. This author records thirteen cases of recovery, six being of the furious form, out of one hundred and fifty-nine inoculations to the dog. The immunity which followed these recoveries was still in force after five years.

Most generally, immunity is observed after inoculations which are not followed by rabic symptoms; it may be obtained in various ways. Pasteur obtained it in a dog by subcutaneous inoculation of street virus in large amount; this immunity, however, is not the rule, and in some cases rabies itself appears.

Hoegyes obtained immunity by injections of fixed virus diluted in 0·7 per cent aqueous salt solution. He made six inoculations in the dog with virus diluted to 1 to 5,000, 1 to 2,000, 1 to 500, 1 to 100, and 1 to 10. The animal showed no sickness and contracted a solid immunity against bites and diverse experimental inoculations.

Immunity was obtained by Pasteur by inoculating the dog with virus more or less attenuated in the ape.

The same author has conferred immunity on the dog by subcutaneous inoculation of virus attenuated by the method of desiccation above described. The first inoculation is made with the cord of a rabbit desiccated during fourteen days, that is, one which had lost virulence for the rabbit seven days before; on the following days inoculations were made with cords which had been subjected to a less lengthened desiccation and, finally, with the virulent cord. The latter was inserted under the skin, in the blood, or in the cranial cavity; some of these vaccinated rabbits were subjected to bites of rabid dogs; these different virulent inoculations produced no effects, and the persistence of the refractory state was verified several years later.

Pasteurian vaccination.—The period of incubation of of rabies after bites being quite lengthy, Pasteur conceived the idea of vaccinating, during this period, the subject which had received the bite, with the view of preventing the invasion of the disease. Experience has justified this illustrious *savant* in his belief. He has succeeded, by means of a series of dried cords, in vaccinating bitten dogs and even dogs which had been inoculated by trephining. It is evident that such preventive inoculations should be made as soon as possible in order that the refractory state be established before the active virus attains the nerve centers. The results obtained in animals have served as a basis for the method of preventing rabies after bites in the human species, a method which is now applied in several establishments, and the beneficial results of which are no longer contested. Statistics show that the mortality from rabies in man has fallen from 16 per cent to 0·67 per cent.

Prevention of rabies after bites in ruminants.—M. Galtier has shown that the intra-venous injection of ruminants with the virus of street rabies does not communicate the disease but confers immunity. This author first, and later MM. Nocard and Roux, showed the efficacy of this method in the protection of ruminants recently bitten by rabid dogs. An emulsion is made with virulent cord in sterilized water, strained through a linen cloth and injected into the jugular or ear vein. The injection, which should be of large amount, is not attended with danger.

*Typhoid fever of the horse.**

Under the title of the typhoid affections several diseases were formally included which, clinically, had not been sufficiently differentiated. Our knowledge of these diseases is now somewhat more extended through the labors of Schütz, Perroncito, Chantemesse and Delamotte, Cadéac, Galtier and Violet, etc. Many points are yet to be elucidated, but two diseases at least have been described which should not be included in the group of typhoid affections. These are pneumonia or contagious pleuro-pneumonia, and pneumo-enteritis of fodders which will be considered later.

Equine typhoid fever is especially characterized by stupor and great depression of the nervous and muscular systems of the affected animals; they are weak and unsteady on their feet, hold the head low, and move with a staggering gait as if under the influence of a profound intoxication or narcosis; the eyes are weeping, half-closed, and often inflamed; the conjunctival mucous membrane is infiltrated, blood-shot,

* [The term "typhoid fever," or "typhoid affections" of the horse appears to include the same diseases or complications of one disease which in England and America are described as Influenza, Epizoötic cellulitis, etc.—D.]

and of a yellow tinge which may become yellowish-red when conjunctivitis supervenes.

The fever of invasion is followed by diverse localizations; generally the digestive symptoms predominate; the tongue is dry and more or less thickly coated, and symptoms of gastro-enteritis quickly appear. When the lungs become affected this complication occurs always several days after the beginning of the disease. The changes in this organ also differ absolutely from those which characterize infectious pneumonia. Typhoid fever predisposes to passive congestions; hence the lung becomes œdematous rather than hepatized and tubular breathing over the affected region is never heard, as in the case of pneumonia. This tendency to venous hyperæmia shows itself in the limbs by œdematous engorgements.

All attempts to transmit the typhoid disease have been ineffective; the horse, ass, dog, and rabbit failed to contract the disease by the different methods of inoculation in common use. Equine typhoid, however, conducts itself like a contagious disease; unlike contagious pneumonia it is polymorphic, showing itself sometimes as an enteritis, sometimes as a carditis, and sometimes as a disease of the lungs, etc., whilst in a stable in which pneumonia prevails, all the affected horses show the latter lesion from the first.

The formerly mooted question as to the identity of equine and human typhoid fever is now settled; Eberth's bacillus does not occur in the equine disease and its inoculation to the horse remains without effect.

*Contagious pneumonias of the horse.**

The results obtained by different investigators regarding contagious equine pneumonia are far from being concordant. Nevertheless, it has been well established that we must separate from the typhoid affections one or more diseases which especially involve the lung. Whilst in typhoid fever the pulmonary localization is delayed, in infectious pneumonia the fever of invasion is of short duration and the disease localizes itself in the lung from the first. The early signs of depression in pneumonia are not to be compared with the state of prostration of animals affected with typhoid fever, and these symptoms subside when hepatization is accomplished; then the animals recover their accustomed liveliness. The conjunctival mucous membrane has a saffron red tint, the eye is bright and well opened. A rusty discharge from the nose appears during the first stage.

The pulmonary inflammation may be lobar, occasionally lobular, or it may be complicated with pleurisy (pleuro-pneumonia); other complications occasionally supervene, affecting the kidney, synovial membranes, articulations, the heart and its serous coverings, the meninges, nerve centers, gastro-intestinal apparatus, etc.

Galtier and Violet assert that the intestine is often affected along with the lung, even as a primary lesion (pneumo-enteritis), and that the disease also affects,

* [In English veterinary works referred to as a complication of influenza. *Ger.* Brustseuche.—D.]

secondarily, the liver, spleen (it becomes enlarged and uneven on the surface), the bladder, articulations, tendinous sheaths, muscles, and keratogenous membrane.

Schütz has described, as the cause of this disease, an ovoid microbe, often associated in pairs and which possesses a capsule comparable with that of the pneumococcus of man. Perroncito has also observed a capsule, but states that it does not react to staining agents like that of the pneumococcus; further, the microbe studied by Perroncito kills the rabbit and the guinea pig, whilst the pneumococcus is inactive for the latter. Chantemesse and Delamotte attribute the disease to a streptococcus. Galtier and Violet describe two different microbes which gave two similar diseases: a streptococcus and a diplococcus (*streptococcus et diplococcus pneumo-enteritis equi*). Cadéac found in all cases only one micrococcus, often grouped in pairs, sometimes in chains.

In some investigations of our own, (1) we have found

(1) In several horses affected with pneumonia we have found the streptococcus already described by Chantemesse and Delamotte and studied by Galtier. The elements of the chains are stained by the method of Gram and Weigert; cultures in bouillon produce a flocculent precipitate; gelatin is not fluidified. Inoculation in the rabbit causes a rapid emaciation and a severe diarrhœa, which results in death; the latter is often preceded by pulmonary hemorrhage. At the autopsy the blood is dark, little plastic, and the serum deeply tinged by the coloring matter of the dissolved corpuscles; the cavities of the peritoneum, pleura and pericardium always contain an abnormal quantity of reddish colored serum. The intestines are highly inflamed; Peyer's patches are injected, sometimes studded with petechiæ; the liver is discolored and tumefied; the spleen is enlarged, uneven, dark and friable; the kidneys show congestion and extravasation; the pleura

the streptococcus of Chantemesse and Delamotte. The microbe of Schütz and those of Galtier and Violet do not take the Gram, whilst all the others do. Most of them are facultative anaërobes. All are pathogenic for the rabbit, which contracts a rapidly fatal disease: the blood is decomposed and the coloring matter of the corpuscles, thus set free, passes into the serum, giving to the latter a red tinge, which is communicated to the liquids of the serous membranes and to the viscera in contact with these liquids; the

sometimes contains a fibrinous exudate; the lung is much congested, this condition being sometimes complicated with interstitial hemorrhages; the blood often escapes into the bronchi unchanged; once only was hepatization present and that in a rabbit which had survived five days.

The guinea pig is a very unreliable reagent for the streptococcus, whilst the dog is unaffected by it.

Inoculation of 5 cc. of a culture into the lung of a glandered horse caused a very intense febrile reaction; the temperature rose from 38·5° to 40·1° and remained at this figure during several days, whilst the animal showed complete inappetence and great prostration; it usually kept the recumbent position and had to be assisted to its feet. It was impossible to detect any symptoms of pulmonary disease. At the autopsy, made ten days later, nothing was found at the place of inoculation, but four pneumonic centers existed on the lower border of the lobes of the organ. Three of these had the dimensions of a five franc piece, the fourth that of a child's hand. At these places the lung was very consistent, of a deep brown color, and manifestly hepatized. Two other horses received an injection of a five days old culture in the lung. At the autopsy, made on the following day in one case, and on the third day in the other, the place of inoculation showed an inflammatory focus large as a man's fist; the hepatized lung was dotted with very fine hemorrhagic points and the visceral pleura infiltrated with plastic serosity and notably thickened. In these lesions the streptococcus of the cultures was found. None of these last cases showed any febrile reaction.

spleen is much enlarged. The diplococcus pneumo-enteritis equi has a less pronounced dissolving action on the blood corpuscles. Inoculation to the horse causes pneumonia; this disease develops as a consequence of injection into the lungs, trachea, or circulation, but only rarely as a result of ingestion.

According to Galtier and Violet, the disease originates when horses are fed on forage of bad quality, such as soiled, moldy, or rusty hay, and imperfectly harvested and damaged grain. These damaged foods are the bearers of the germs of the disease and it is by their intermediation that these germs obtain entry to the system. The dust rising from the fodder enters the respiratory passages, where the microbes which it carries along act directly upon the lung. The disease has, in fact, been produced in the horse by inoculation of the products obtained by maceration of suspected foods.

The disease can then be transmitted from one animal to another by the excrements and nasal discharge; contagion, according to Galtier, plays an accessory part; the diffusion of the cause suffices to explain the enzootic and even epizootic character of the affection. Its transmission, however, can not be doubted.

Contagious pleuro-pneumonia of cattle.

Microbe.—M. Arloing found that the serosity which flows from the surface of a section through a diseased lung is very poor in microbes and that the majority of the culture bulbs inoculated with a small quantity of this serosity remain sterile. To obtain fertile cultures a large quantity of this serosity must be transferred to the culture medium, or the inoculation made with the material obtained by scraping the surface of the section. By sowing the latter product on gelatin M. Arloing isolated four different species, one being a bacillus (*pneumo-bacillus liquefaciens bovis*) which rapidly fluidifies gelatin, and the other three micrococci. Of the latter, one produces white colonies resembling drops from a wax candle, the second gives whitish colonies which become wrinkled on aging, and the third, colonies which take an orange yellow tint. M. Arloing attributes the disease to the *pneumo-bacillus liquefaciens bovis*. The four microbes inoculated separately under the skin of a steer give an inflammatory tumefaction which disappears in five or six days; the largest tumefaction is caused by the bacillus, and if several successive generations are inoculated it ultimately happens that the bacillus alone produces a local reaction. Moreover, the bacillus alone is constantly present in diseased lungs. M. Arloing noticed that the isolated effects of the microbes which he had cultivated resembled only in a remote way those produced by fresh serosity, but he observed, also, that this last becomes more active in passing through the cellular tissue of healthy cattle. By taking the microbes from this reinforced serosity he obtained more virulent cultures of the pneumo-bacillus; 4 cc. of culture injected into the lung of one steer and 20 cc. injected into the veins of another produced the specific lesions of pleuro-pneumonia. This author isolated from cultures a soluble substances which possesses remarkable phlogogenic properties, and which, of itself alone, reproduces the characteristic inflammatory engorgements of the hypodermic inoculation of active virus. He attributes to this substance the inflammations which sometimes occur, in the course of the natural disease, in points remote from the thorax.

Whatever opinion may be held regarding these investigations, the virulent agent of pleuro-pneumonia exists in the serosity which flows in abundance when a diseased lung is incised. Inoculation of this liquid produces the following effects:

In the subcutaneous cellular tissue a more or less intense inflammatory engorgement results; in regions in which the connective tissue is loose and abundant this reaction often assumes a severe, progressive and occasionally gangrenous character, and leads to death; the general effects of the virus are indicated by a febrile reaction of greater or less intensity. Pleuro-pneumonia is almost never observed as a consequence of subcutaneous insertions. But the latter confers immunity against later inoculations.

The injection of the virus into the veins gives no more characteristic pulmonary lesions; it confers immunity without any local manifestation, unless some of the virus should fall on the perivascular cellular tissue, in which case a dangerous tumor results. M. Thiernesse, however, observed pleuro-pneumonia after an injection of 35 grams in the jugular.

Vaccination.—In 1852 Willems recommended preventive inoculation against this disease; his method was put in practice in different places and the preventive action of Willemsian inoculation placed beyond doubt.

The inoculation may be performed in three principal ways: by dermic, hypodermic, and intravenous insertion. Dermic insertion gives uncertain results, the virulent matter requiring to be brought into the cellular tissue in order to act efficaciously. On the other hand, hypodermic injection supplies the best conditions for the evolution of the virus. For the reception of the pulmonary serosity the cellular tissue of the tail should be selected in preference to all other regions; as has already been said the consequences of inoculation are often very severe in regions where the connective tissue is loose and abundant. In the tissue of the tail, one or two drops of the serosity are sufficient. It is well to make a second vaccination a few weeks later; this may be made in an interdicted region although this procedure can not be recommended, especially when the reaction to the first inoculation has been insufficient.

With the view of obviating the accidents which occasionally follow the caudal inoculation some experimenters, including Professor-Director Degive, have had recourse to intravenous injection; this method procures a solid immunity, but requires special care in order to avoid accidental contact of the virus with the cellular tissue which surrounds the vein, an accident which may result in a serious, perhaps fatal, engorgement. Larger doses may be introduced into the veins than into the cellular tissue.

For the practice of Willemsian inoculation it is essential that fresh virus be employed. This is obtained by making a clean incision of a diseased lung with an aseptic knife and collecting the serosity which flows spontaneously from regions in which the inflammation is most recent. As such a lung is not always at hand efforts have been made to preserve the virus.

M. Laquerrière showed that in a frozen lung the virus remains intact for one year, at least.

M. Nocard has quite recently recommended the preservation of the virus by the addition of half a volume of a five per cent solution of carbolic acid and half a volume of pure neutral glycerin. This mixture retains its virulence for months.

For the purpose of procuring pure virus some have advised its cultivation by direct inoculation in the cellular tissue of the calf and collecting the serosity of the inflammatory engorgement.

Others again have sought for a means of mitigating the effects of the natural virus in order to diminish the accidents which result from its use; Pasteur has shown that it is preserved for six weeks in sealed tubes, but, at the same time, becomes so attenuated that it can be inoculated in an interdicted region without exciting fatal lesions. It has also been proposed to dilute the virus in water. Dilutions of 1 to 50, 1 to 100, and 1 to 500 have yet sufficient activity to produce considerable inflammatory reactions and, therefore, to confer immunity.

Septic pleuro-pneumonia of calves.

This disease occurs in an enzootic form on certain farms, attacking and quickly killing calves while still quite young. The lesions by which it is characterized have some analogy with those of contagious pleuro-pneumonia of larger cattle, but the thickening of the connective tissue septa is less marked and the flow of serosity less abundant, whilst the individual pulmonary lobules have not the uniform color seen in the latter disease. We have observed that the pulmonary lesions begin, as in pleuro-pneumonia, in the interlobular connective tissue and, in the peripheral lobules, progress toward the center. Poels has noted, in calves affected with this disease, the frequency of sero-fibrinous exudates and pleuritic adhesions, lesions which have also been recorded by M. Vanden Maeghdenbergh. He also mentions the occasional occurrence of inflammations of the pericardium, liver, kidneys, stomach and intestine.

Microscopical examination shows, in the lung and in the muco-pus of the bronchi, the presence of ovoid microbes with rounded ends, measuring from 1μ to $1{\cdot}5\mu$ in length by $0{\cdot}5\mu$ in thickness, easily stained by the aniline colors, but not stained by the Gram or Weigert methods. When stained with a very dilute aqueous solution of gentian violet they fix the color at their extremities, whilst the center remains clear. They are motile, vegetate rapidly in bouillon and on solid media, and are pathogenic for various species of animals.

The rabbit dies in twenty-four to forty-eight hours after subcutaneous inoculation or ingestion of cultures or virulent products. In the same animal intra-pulmonary inoculation of a drop of culture produces a pneumonia. Death is a little later in the guinea pig than in the rabbit.

Two calves, aged thirteen days and eight weeks, and a one-year-old heifer, were inoculated by Poels, the first in the right pleura, the second in the trachea, and the third in the lung. They died after 20, 54, and 66 hours, respectively, with lesions of septic pleuro-pneumonia. A pig also took the disease by pulmonary injection; the sheep and the dog are refractory (Poels). The microbe is described as a facultative parasite capable of living in the soil, which fact, according to Poels, explains the persistence of the disease on an infected farm. M. Galtier thinks that the latter ought to be attributed to the mothers (see page 325). However that may be, its transmission from calf to calf must be admitted; the virulent germs are spread around with the expectorations and perhaps with the excrements (Poels claims to have found the germs in all the organs). These microbes, and especially those of the nasal discharge, contaminate the vessels in which milk is fed to the diseased, and the same vessels are then used for other animals.*

The microbe of this pleuro-pneumonia of calves belongs to the great class of ovoid bacteria showing

*[In the diarrhœa or dysentery of young calves known as "white scours" (*Ger.* Kälberruhr) Jensen (1892) discovered in the blood, spleen, liver, kidneys and lungs, as well as in the mucous membrane of the intestine, oval bacteria, isolated or associated in pairs or short chains, staining at the extremities only, and easily cultivated in the various artificial media. Bouillon cultures fed to young calves in doses of 5 cc. produced the characteristic diarrhœa and death in from one to two days. From the contents of the intestine of healthy calves he isolated apparently the same germ but found it destitute of pathogenic properties. Jensen came to the conclusion that the microbe is a facultative parasite, a usually harmless inhabitant of the intestinal canal but one which under certain abnormal conditions of the intestine (perhaps attributable to the diet) acquires pathogenic properties which become increased by subsequent passage from calf to calf.—D.]

a clear central space, which are met with in a series of diseases. Hueppe groups these affections under the name of *hemorrhagic septicæmias* and attributes them all to the same germ. These diseases are: Koch's rabbit septicæmia, fowl cholera, duck cholera, parrot disease, infectious pneumonia of the pig, pneumo-enteritis of the pig, pneumo-enteritis of sheep, epizootic of wild game (Wildseuche), epizootic of ferrets, disease (barbone) of buffaloes, etc. It would be superfluous to insist on the non-identity of the germs of these different affections, but their close relationship can not be denied. The morphological characters of the germs are almost identical; all take the stain only at their extremities, leaving a clear space in the middle; they do not stain by the Gram method, do not fluidify gelatin, and are pathogenic for rabbits. It is probable that several of the diseases mentioned above are due to one and the same microbe (fowl cholera and rabbit septicæmia, for example) and that all these micro-organisms represent varieties of one fundamental species.*

* [Billings found in specimens of the blood and organs of cattle which had died from the so-called "corn-stalk" or "corn-fodder disease" of the Western States an organism presenting great morphological resemblance to the germ of "swine plague" (hog cholera) but showing slight differences in its growth on the various culture media. It belongs to the group of ovoid, bi-polar staining organisms and, according to this author, is the cause of the disease above named. The germ was found to be pathogenic for mice, rabbits and guinea pigs; subcutaneous inoculation of a steer gave rise to fever, pneumonia and pleurisy, and great emaciation, terminating in recovery. From the original communications on this subject we learn that the germ is identical with a bacterium described by Burril as the cause of a disease of corn (maize) (*Ills. Univ. Exper. Sta. Bul.* 6), and the cattle become infected by feeding on the leaves of this diseased corn. Horses are also said

Epizootic abortion.

Epizootic abortion is a contagious disease most frequently seen in the cow but also noticed in the ewe, she-goat, and even in the mare. The abortion occurs at all periods of gestation after the third month; in the same animal it occurs at a later stage in each succeeding year and at length allows the fœtus to be carried till term, if, indeed, the cow, after a first abortion, has not become sterile. The calf is most frequently still-born; in some cases it is born alive but its health is precarious; shortly after birth it emits a peculiar lowing; after the third day it is attacked with diarrhœa and death quickly follows.

According to Nocard the disease takes its origin in diverse germs met with in the uterus of the animals which abort, germs which are never found in healthy

to die from the same disease (*Neb. Agric. Exper. Station, Bulletins* 7, 8, 9, 10).

Nocard studied an "infectious broncho-pneumonia" which was observed in a number of recently imported American cattle (from Virginia, Indiana and Illinois). The lesions in the lung somewhat resembled those of contagious pleuro-pneumonia. The muco-pus of the bronchi, the hepatized lung tissue, and the serosity contained a short, ovoid, motile bacterium, apparently in pure culture. It measured barely 1μ in length by 0·3μ to 0·4μ in thickness, stained with aqueous solutions of fuchsin and methylene blue, leaving a clear unstained central space; not stained by the methods of Gram or Weigert. The germ was found to be pathogenic for the mouse, rabbit, guinea pig and pigeon, which die from subcutaneous inoculation; inoculated in the lung of calves and sheep it occasions a fatally-ending exudative broncho-pneumonia. Nocard believes this disease to be identical with the "corn fodder disease" of Billings. (*Recueil de Méd. Vét.* Aug., 1891.—D.]

animals. These germs, among which is often found a micrococcus, isolated or in chains, are also present in the amniotic fluid, in the digestive canal of the aborted calves, as well as in the substance of the medulla oblongata of those individuals which during life gave utterance to the peculiar lowing sound just referred to. Nocard thinks that these germs give rise to a disease of the fœtus and its envelopes, the mother remaining healthy. He explains the repeated abortion by the persistence of these germs in the womb, and the sterility by the acid reaction which they produce in the uterine secretions.

The invasion of a stable by this disease generally coincides with the introduction of a pregnant infected cow; abortion then occurs annually in a certain number of animals in this stable. The lengthy period of incubation of the disease implies an early infection.

Galtier thinks that the disease is due to a general infection of the mother, which communicates the disease to the fœtus. Calves which are not prematurely expelled will still harbor the germs, and these germs produce the pneumo-enteritis which occasions such ravages in these animals.

In short, epizootic abortion appears to depend on multiple causes which are yet to be discovered, but in all cases the method of treatment laid down by Nocard is to be recommended.(1)

(1) The prophylactic measures recommended are as follows:

1. Each week the floor of the stable should be scraped, thoroughly cleansed, and sprinkled with a solution of sulphate of copper: 40 grams to the litre.

2. Each week from the date of conception the vagina of the

Contagious mammitis of milch cows.

MM. Nocard and Mollereau have seen and described a special form of mammitis occurring in milch cows and readily passing from one animal to another. It appears in the form of indurated lumps which commence at the base of the teat, gradually increase in size, and may sooner or later invade the whole organ. The milk is diminished in quantity, becomes acid in reaction, and soon coagulates—often as soon as it is drawn from the udder; it is often mixed with pus, and grumous; sometimes it exhales an offensive odor. These characters it communicates to good milk with which it may be mixed.

Microbe.—This is a rounded or ovoid micrococcus; it measures 1·25μ in length by 1μ in thickness and forms long straight or sinuous chains. It frequently appears bi-lobed, in way of division. It is aëro-

pregnant cows should be thoroughly injected by means of a large syringe, with the following tepid solution:

Distilled water,	. . .	20 liters.
36 per cent alcohol } āā Glycerin, }	. . .	100 grams.
Bichloride of mercury,	. .	·10 grams.

3. Each week, at the time of grooming, the vulva, anus, and lower surface of the tail of all the pregnant cows should be carefully washed with a sponge saturated with the same tepid solution.

4. If a cow should abort it will be necessary to remove the placenta by hand, to destroy the fœtus and after-birth by fire or boiling water, and to irrigate the uterus by means of a long tube introduced to the bottom of the cavity, with eight or ten liters of the tepid solution indicated above but containing only half the proportion of sublimate.

anaërobic. These characteristic chains are found in the milk and in the wall of the excretory ducts.

Action of physical and chemical agents.—The growth of the microbe in cultures is checked by a trace of boric acid; MM. Nocard and Mollereau, taking advantage of this peculiarity, injected on several occasions at intervals of eight days 100 grams of a tepid four per cent aqueous solution of boric acid into the teats of the affected udders, the injection naturally being made immediately after milking. The parasite is also destroyed by a three per cent solution of carbolic acid; the authors recommend this solution for washing the hands of those who undertake the milking. By these measures they succeeded in arresting the extension of the disease.

Cultures.—Alkaline bouillon with the addition of sugar or glycerin forms a medium well suited for its growth; at 35° it forms a mass of very long chains which sometimes become agglomerated in silky flakes which after several days are deposited; the reaction becomes acid in twenty-four to forty-eight hours; if chalk be added to the bouillon so as to neutralize the acid as it is produced, the culture is more vigorous and retains its vitality longer. Crystals of lactate of lime are often found at the bottom of the bulb. The growth of this microbe, therefore, gives rise to the lactic fermentation; it can not, however, be identified with the lactic bacillus. Cultures almost invariably die after a few weeks. The microbe grows also on the different solid media.

Experimental inoculations.—Inoculation of pure cultures into the teat reproduced the disease in the cow and the goat; from the first day the inoculated udders supplied a milk very rich in streptococci; that from the cow quickly showed an acid reaction and became clotted; the udder finally became inflamed. Inoculation in the mamma of a nursing bitch remained without effect; the dog, cat, rabbit, and guinea pig were also unaffected by intra-venous and intra-peritoneal inoculations.

Etiology.—The disease is communicated through the intermediation of those who have charge of the milking of the cows, their hands being soiled with the diseased milk. When the latter is mixed with good milk this also takes the same characters. Contagion to the cows must, therefore, be prevented by disinfection of the hands, and contagion to the normal milk by keeping the milk from the diseased cows in separate vessels. Such milk is, moreover, unfit for human consumption.

Besides this disease, remarkable on account of its extreme contagiousness, the udder of the cow is subject to various microbic lesions. M. Lucet especially called attention to infectious forms of mammitis due to an external cause; in a series of cases of acute mammitis he found one or several germs; these were sometimes micrococci, sometimes bacilli, and sometimes both together. Penetration of the germs most frequently occurs through solutions of continuity of the integument of the udder. The microbe being different in different cases, we can readily understand that the severity of the disease will be very variable.

Among the number of infectious forms of mammitis with internal cause tubercular mammitis should be especially mentioned.

Gangrenous mammitis of milch ewes.

This disease, also called *mal de pis*, *araignée*, quickly kills ewes which are attacked by it; it is due, according to the researches of M. Nocard, to a very fine micrococcus, measuring 0·2μ in diameter, and associated in groups of four or more, never in chains. It is stained by the method of Gram, is aëro-anaërobic, and communicates to bouillon and to milk an acid reaction, coagulating the latter in twenty-four hours. Cultures retain their virulence only when renewed every day. When a culture is inoculated into the teat of a ewe it produces a rapidly fatal mammitis. The goat is refractory. The rabbit contracts an abscess from which it quickly recovers. The dog, cat, and guinea pig show only a local œdema.

Diseases of milk.

Milk just withdrawn from the udder is free from germs, but after its extraction it may quickly become infected and undergo a series of deteriorating changes. The most important of these deteriorations will here be briefly described.

Curdled milk.—The curdling of milk results from the lactic fermentation of milk sugar, the acid produced then bringing about the coagulation of the casein.

The usual cause of this fermentation is the *bacterium lactis.* This is a short, non-motile rod, measuring in length 1μ to 3μ by 0·6μ in thickness, most frequently isolated but occasionally arranged in series, and capable of spore formation.

Other germs give rise to the same deterioration of milk, among these being the cocci of suppuration, erysipelas, contagious mammitis of the cow, etc. The infection occasionally originates in the udder (mammitis), more frequently, however, after the milk is withdrawn. The infection is facilitated by lack of cleanliness in the stables, dairy utensils, etc.

The diseased milk coagulates more or less quickly according to circumstances; the process is hastened by heat. The cream separates imperfectly from such milk and the agglomeration of the butter globules is difficult.

Putrid milk.—Milk, like all organic liquids, readily putrefies. Various putrefactive germs are concerned in this process, but those which are most constant are the *bacterium termo* and *lineola.*

The former is represented by short motile rods, measuring 1·4μ by 0·7μ. The second species consists of large cylindrical, motile rods, measuring 3μ to 5μ by 1·5μ.

Infection results from lack of cleanliness; the infected milk quickly putrefies, with the production of putrid gas in its substance. At the same time the cream takes a yellow color and a bitter or rancid taste, and butter can no longer be obtained from it; it then gradually disappears.

Viscous milk.—In this form of deterioration, which is observed one or two days after milking, the milk is not readily coagulable and the cream separates imperfectly; butter is obtained with difficulty and has a disagreeable taste.

The disease is due to rounded elements 1μ in diameter, isolated or associated in the form of chains, and transforming the lactose into a mucilaginous substance which gives to the milk its peculiar consistence.

Blue milk.—It sometimes happens that milk, from twenty to thirty-six hours old, shows on its surface small light-blue patches which, later, take an indigo-blue shade. These patches increase in area as well as in depth although never exceeding the thickness of the cream. Such milk is highly subject to change; it quickly becomes acid, coagulates and then putrefies.

The causative agent of blue milk is the *bacillus cyanogenus.* This is a motile rod, 2μ to 4μ long by $0{\cdot}5\mu$ thick, commonly isolated but sometimes united in a zooglœa. The spore which it produces is a little larger than the bacillus giving to the latter a club or spindle-shaped appearance. In certain artificial media it assumes very diverse involution forms—balloon and ribbon shapes, etc. Grown on the gelatin plate it forms, after two days, small whitish spots which soon extend over the whole surface giving it a bluish color. Stab cultures in gelatin show whitish colonies on the surface and steel blue in the depth.

When inoculated to milk it increases the alkalinity of the latter and the layer of cream becomes slate colored, this tint turning to blue on the addition of acid. If the milk has not been sterilized the lactic acid fermentation which goes on at the same time supplies the acid by which the blue color is produced.

The germ multiplies in albuminous solutions containing lactate of ammonia, without producing the coloring principle.

This coloring matter is most abundantly produced at about 20°, in less amount at 25°, and not at all at 37°. The pigment is a soluble substance which gives a red reaction with potash, violet with ammonia. Inoculation of this germ to animals produces no result.

The infection of milk takes place after its withdrawal from the udder and is contingent upon a previous infection of the dairy or of the stable.

Cheese made from infected milk may also show this blue color, but the latter changes more or less to green in consequence of the yellow color which cheese assumes in aging.

Red milk.—Two species of germs are capable of imparting a red color to milk. One, the *micrococcus prodigiosus*, is elliptical, motile, and forms on potato an abundant slimy growth of blood-red color. The infected milk shows on the surface a pellicle of a more or less deep red color, the deeper layers remaining unaltered. Another, the *bacterium lactis érythrogènes*, is a very short non-motile rod. In milk it slowly precipitates the casein and the whole mass assumes a blood-red color.

Red milk is of much rarer occurrence than blue milk.

Yellow milk.—This deterioration occurs especially on boiled milk; golden yellow patches appear on the milk which, at the same time, coagulates and becomes alkaline. The change is produced by the growth of the *bacillus synxanthus*, a slender, very mobile rod.

Bacterial hæmoglobinuria of cattle.

This acute febrile disease prevails in an endemic

form in certain marshy regions of Roumania, where it causes the loss of large numbers of oxen. Cows are less susceptible while calves seem to be refractory. It is characterized by the presence of albumen and hæmoglobin in the urine; the latter is red colored but does not contain blood corpuscles. The autopsy reveals the presence of interstitial extravasations and ulcers of the fourth stomach and duodenum. The tissue surrounding the kidneys is infiltrated with blood and serosity, the kidneys are friable and dark red in color and their pelvic mucous membrane ecchymosed; the bladder is filled with red colored urine. The liver is tumefied and discolored, the spleen enlarged and darkened and the pulp diffluent.

Babès has discovered in this disease a rounded microbe 0.5μ in diameter, usually arranged in pairs, sometimes in tetrads; it is decolorized by the Gram stain. It is found in the blood adhering to the red blood corpuscles or situated in their interior, but more especially in the serosity of the hemorrhagic œdemas and in the vessels of the kidney. It also exists in the vessels of the intestinal ulcers. The invaded red blood corpuscles are more or less altered.

The rabbit, by inoculation of the blood or œdematous liquid, as well as by ingestion of the products of the disease or its cultures, contracts a general disease which often terminates fatally. In the ox, the introduction into the veins or connective tissue of a considerable quantity of blood or of juice expressed from the kidneys, reproduces the typical disease with hæmoglobinuria.*

* [In many respects similar to this hæmoglobinuria of cattle and to a closely allied disease of sheep ("Carceag") also investigated

Distemper of young dogs.

The contagiousness of this disease is well established although our knowledge of the germs which

by Babès is the disease of cattle known in America as "Texas fever." It occurs in an acute and in a mild form, the latter being in a sense endemic in the southern United States, whilst the acute form prevails in northern cattle which have been imported to these regions or which have been exposed to the infection brought north by cattle from the infected districts. The chronic or mild type, as it occurs in southern cattle, generally passes unperceived. Mild forms of the disease are also frequent among northern cattle, especially when they receive the infection during the cooler weather of autumn. The acute disease (which occurs in the northern States only during the hot weather of summer) is generally fatal; in non-fatal cases it is sometimes followed by a prolonged period of unthrift and debility. Calves are less susceptible than adult animals and after one or more attacks acquire a certain degree of immunity; to this cause is attributed the comparative immunity possessed by cattle native to the southern States. It appears, however, that this immunity is not permanent and may be lost when such cattle are kept for several summers in non-infected regions. The most noticeable symptoms of the acute disease are those of an intense continuous fever with the passage, in most cases, of red or dark-red colored urine; unsteady gait and muscular tremors in the neck and limbs occur toward the last stage; occasionally symptoms of delirium are observed.

Pathological anatomy.—The most important lesions observed at the autopsy are: Injection of the vessels and occasionally patches of extravasation in the subcutaneous connective tissue; blood thin and watery (Smith), or frequently of normal appearance; lungs normal or discolored by congestive patches; heart muscle congested, points of extravasation on the pericardium and endocardium; spleen enlarged, its capsule streaked and mottled by the injected vessels, pulp dark red and diffluent; liver generally much enlarged and darkened from blood congestion or light yellow in color from extreme engorgement with bile; gall bladder full of thick, dark, grumous bile; kidneys generally congested, determine it is as yet incomplete; some authors have described a bacillus; others, a bacillus and a coccus;

often showing irregular discolored patches beneath the capsule which itself is non-adherent, points of extravasation in the pelvic mucous membrane; in many cases extensive effusion of bloody serosity in the fat and connective tissue surrounding one or both kidneys; the urine contained in the bladder generally dark red in color free from blood corpuscles but strongly albuminous, sometimes of normal appearance; congestion of the mucous folds of the fourth stomach; more or less hyperæmia and extravasation in the walls of the small intestine have also been noticed.

Etiology.—Texas fever in its acute and best known form occurs under two conditions, 1st, when northern cattle are shipped into southern infected regions, and second, when cattle from permanently infected regions, brought north during the summer months, infect pastures in which susceptible cattle are subsequently allowed to graze. In the first case the disease may, and frequently does, appear two weeks after exposure to infection, in the second a lapse of six weeks or more intervenes between the arrival of the infection-bearing cattle and the outbreak of the disease. These differences are attributable to the method by which the infection is conveyed into the bodies of susceptible animals. The investigations of Smith and Kilborne (afterward repeated with the same results by others) have shown that this takes place through the intermediation of cattle ticks (*Boophilus bovis*, Curtice), the progeny of those adherent to the skin of the southern cattle. The lapse of time between the arrival of these cattle and the outbreak of the disease represents the time required for the incubation of the next generation of these parasites along with the period of incubation of the disease. In the case of cattle shipped into permanently infected districts their invasion by ticks and consequent infection may begin at once. It is possible also that in permanently infected districts infection may occur by other means. Later generations of ticks which come to development on susceptible cattle in the North are also capable of communicating the disease. Horses, which may also be invaded by these parasites, do not obtain the disease, and the progeny of ticks which have developed on these animals have not been shown to be dangerous for cattle.

Experimental inoculations.—Positive results have been obtained and others, again, attribute the disease to a micro-coccus. We have found as the result of our own

by Smith by intra-venous and subcutaneous injection of susceptible cattle with the blood of animals suffering from the disease as well as with the blood of apparently healthy southern cattle. No other animal species has been found to be susceptible. Direct contagion seems, however, rarely or never to occur under natural conditions.

Microbe.—Cover glass preparations from the blood, spleen, liver, kidney, and heart muscle of cattle which have died from acute Texas fever, show the presence of a rounded or somewhat ovoid, or pyriform (Smith) body, isolated, in pairs, or occasionally three or four, within a certain proportion of the red blood corpuscles. They are usually extremely abundant in the juice of the kidney (where they also occur between the cellular elements), less so in the liver and spleen, and still less in blood from the large vessels or heart. These stain readily with aqueous solutions of aniline colors (methylene blue), as well as with hæmatoxylin (Delafield's). Their outline after staining is generally less well defined than bacteria similarly stained; nevertheless their appearance within the unstained disk of the red blood cells is quite characteristic and in the absence of other evidence is of itself sufficient for a diagnosis of the disease.

Fig. 18.

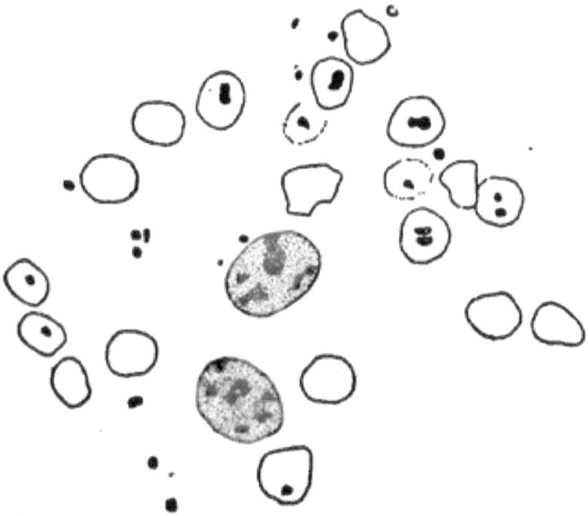

Smear preparation from kidney of ox. Acute Texas Fever. ×1000. (D.)

Culture tests of the blood and organs in cases of Texas fever have, in general, led to negative results. Billings obtained an ovoid "bi-polar" researches, bearing upon four subjects only, that a micrococcus in pure culture is always present in the pustules of the cutaneous form. M. Mathis has recorded in a special work the results of his investigations. He succeeded in demonstrating in the contents of pustules, in the discharge from the nostrils, in the blood and in the tissues, the presence of spherical micrococci, isolated or grouped in pairs, in chains, or in masses; these measure $0{\cdot}1\mu$ to $0{\cdot}3\mu$ in diameter. He cultivated these organisms in bouillon; this medium becomes turbid, then, after fifteen to twenty days, clears again by the deposition of the suspended germs at the bottom of the vessel.

Subcutaneous inoculation of these cultures in susceptible dogs is followed by an œdematous tumefaction, with pustules on the skin in the region of the inoculation; generally there is an elevation of temperature and occasionally generalization of the pustulous eruption, with cough, discharge from the nose, etc., and if young subjects are experimented with, the disease may terminate fatally.

staining bacterium, pathogenic for small animals, which he describes as the cause of the disease. Smith, to whose investigations our knowledge of the etiology of Texas fever is chiefly due, regards the intra-corpuscular body already mentioned, which he discovered in 1889, as the causative agent of this disease. He describes it as occurring in several forms representing different stages in its development; the parasite, according to this author, belongs to the protozoa (Pyrosoma bigeminum). The destruction of red blood corpuscles, the essential characteristic of this disease, is brought about by the direct action of the parasite. The latter exhibits amœboid motion within the corpuscle which it ultimately destroys and then is found in its free stage between the cellular elements. Attempts at cultivation were unsuccessful.—D.]

The experimental disease confers immunity. M. Mathis has found that immunity also follows the natural disease.

Notwithstanding the importance of the results already obtained concerning the etiology of distemper of dogs, there are some obscure points which yet require to be elucidated, such as, for instance, the interesting question of the pathogenesis of those nervous troubles which so frequently complicate the disease.

Phosphorescent meats.

Dead animal matters not unfrequently become phosphorescent. Marine fishes and mollusks are especially liable to be thus affected, while meats are less subject to this change. Phosphorescent meat shows on its surface a coating which is luminous in the dark, and easily removed by scraping.

The cause of this deterioration resides in a bacterium, 1μ in length (*Photobacterium Pflugerii*). This germ grows well on meat and fish, especially at temperatures between 10° and 30°. In presence of oxygen it gives to the culture media a whitish glimmer. The invasion of meat takes place very quickly in summer, one piece of meat being readily infected by another. The multiplication of the germ is rather favored by salting, and it ceases when putrefaction begins. The luminous property is directly connected with the life of the germ. It has not hitherto been found that such meats are poisonous.

In order to complete this subject we should yet have to discuss a number of diseases the microbic nature of which is unquestionable. Thus, cocci and streptococci have been described in the vesicles of aphthous fever, bacilli in rinderpest, micrococci in cow-pox or vaccinia, as well as in enzootic hepatitis of young pigs, etc. But not only is there a lack of unanimity as to the morphology of the germs of these diseases, but the study of their special biological characters has yet entirely to be made, hence, a particular acquaintance with these germs is only of secondary importance to the practitioner.

APPENDIX.

(From Attfield's Chemistry.)

The Metric System of weights and measures is founded on the meter. The engraving represents a pocket folding measure, the tenth part of a meter in length, divided into ten centimeters, and each centimeter into ten millimeters:

Fig. 19.

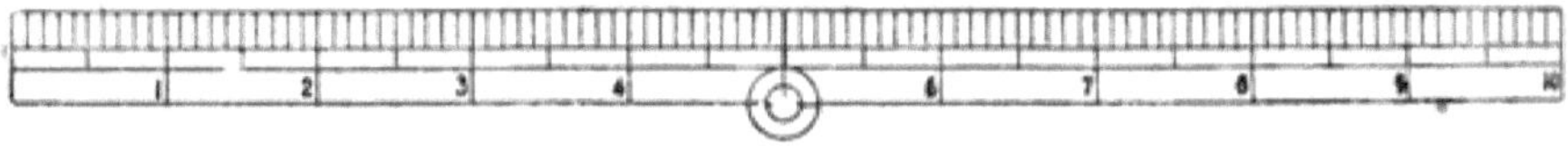

The Decimeter.

The units of the system with their multiples and submultiples are as follows:

UNITS.

Length.—The *Unit of Length* is the METER, derived from the measurement of the quadrant of a meridian of the earth. (Practically, it is the length of certain carefully-preserved bars of metal from which copies have been taken.)

Surface.—The *Unity of Surface* is the ARE, which is the square of ten meters.

Capacity.—The *Unity of Capacity* is the LITER, which is the cube of a tenth part of a meter.

Weight.—The *Unit of Weight* is the GRAMME, which is the weight of that quantity of distilled water, at its maximum density (4 C.), which fills a cube of the one-hundredth part of the meter.

TABLE.

Note.—Multiples are denoted by the Greek words "Deca," Ten, "Hecto," Hundred, "Kilo," Thousand.

Subdivisions, by the Latin words, "Deci," One-tenth, "Centi," One-hundredth, "Milli," One-thousandth.

QUANTITIES.	LENGTH	SURFACE.	CAPACITY.	WEIGHT.
1000	Kilo-meter	......	Kilo-liter	Kilo-gramme
100	Hecto-meter	Hectare	Hecto-liter	Hecto-gramme
10	Deca-meter	......	Deca-liter	Deca-Gramme
1 (Units)	METER	ARE	LITER	GRAMME
.1	Deci-meter	......	Deci-liter	Deci-gramme
.01	Centi-meter	Centiare	Centi-liter	Centi-gramme
.001	Milli-meter	......	Milli-liter	Milli-gramme

Relation of Metric to United States measures of Length, Capacity, and Weight:

1 Meter, ...	39.370432 inches.
1 Decimeter, . . .	3.937043 "
1 Centimeter, . . .	0.393704 "
1 Millimeter, . .	0.039370 "

Approximately, 1 centimeter (cm.) = $\frac{2}{5}$ in.; 1 millimeter (mm.) = $\frac{1}{25}$ in.; or conversely, 1 in. = $2\frac{1}{2}$ cm. = 25 mm.

In micrometry, 1 Micron (μ) = 0.001 mm. = 0.00004 in. = $\frac{1}{25000}$ in.

(Milliliter) 1 cubic centimeter (ccm.) = 16.23 minims.
(Liter) 1,000 cubic centimeters = 33.81 fluidounces.

Approximately, 1 liter = 1 quart.

1 Gramme (weight of 1 ccm. of water at 4° C.) = 15.432 grains.
1 Kilogramme (1,000 grammes) = 15432.350 grains.

Approximately, 1 kilogramme = 2 lbs., *Avoir.*

Thermometric scales.—On the Centigrade (C.) scale the freezing point of water is made zero, and the boiling point 100; on the Fahrenheit (F.) scale the zero is placed 32 degrees below the congealing point of water, the boiling point of which becomes, consequently, 212.

The degrees of one scale are easily converted into those of another if their relations be remembered—namely: 180 (F.), 100 (C.): that is, 18 to 10, or 9 to 5.

Formulæ for the Conversion of Degrees of one Thermometric Scale into those of another:

F = Fahrenheit; C = Centigrade; D = The observed degree.

If above the freezing point of water (32° F.; 0° C.),

F into C, . . . $(D - 32) \div 9 \times 5.$

C into F, . . . $D \div 5 \times 9 + 32.$

If below freezing, but above 0° F (— 17°·77 C.),

F into C, . . $-(32 - D) \div 9 \times 5.$

C into F, . . . $32 - (D \div 5 \times 9.$

If below 0° F (— 17° ·77 C.),

F into C, . . . $-(D + 32) \div 9 \times 5.$

C into F, . $-(D \div 5 \times 9) - 32.$

Equivalents on the Centegrade and Fahrenheit scales:

C	F	C	F	C	F	C	F	C	F
—10	14	25	77	37	98·60	44	111·20	75	167
0	32	28	82·40	38	100·40	45	113	80	176
+5	41	30	86	39	102·20	50	122	90	194
10	50	32	89·60	40	104	55	131	100	212
15	59	34	93·20	41	105·80	60	140	120	248
20	68	35	95	42	107·60	65	149	150	302
22	71·60	36	96·80	43	109·40	70	158	200	392

www.ingramcontent.com/pod-product-compliance
Lightning Source LLC
Chambersburg PA
CBHW030319240426
43673CB00040B/1212
9783337215088